Hamlyn all-colour

John

Aircraft

illustrated by Gerry Palmer

Hamlyn - London
Sun Books - Melbourne

FOREWORD

The twentieth century will surely be remembered in history as the age of flight. After thousands of years of dreaming, and trying to fly with feathered wings like the birds, men learned the secret of powered, sustained and controlled flight when our present century was but four years old.

Since then, flying has progressed from a miracle to a routine that . whisks a million persons from one city to another, one country to another or one continent to another, in armchair comfort, every day of the year. It brings new pleasures to tens of millions annually as they holiday in faraway places. It brings new life to the sick and injured who need urgent medical help, and the promise of more or better food in a hungry world by making barren soil productive and freeing good land from insect pests, weeds and disease. It has made small wars more terrible, yet may have ended the possibility of major war for ever. And it paved the way to the Moon and planets for astronauts and cosmonauts.

The story of practical flying is thus bound up inextricably with the events of our century ... its wars, its constant striving for higher standards of living, its concern for the welfare of others who are less fortunate, and its ceaseless quest for knowledge, even when the cost is almost unbearably high in terms of both money and lives.

Yet, we must never forget those who made the achievement possible, by keeping the dream of flight alive through all the generations that lacked the materials and scientific knowledge to make it come true. We shall never know the names of most of them, but they have their memorials in the stone carvings of winged gods and in the legends of antiquity with which our story of flight begins.

J.W.R.T.

Published by The Hamlyn Publishing Group Limited
London · New York · Sydney · Toronto
Hamlyn House, Feltham, Middlesex, England
In association with Sun Books Pty Ltd Melbourne

Copyright © The Hamlyn Publishing Group Limited 1971

ISBN 0 600 00115 6

Photoset by BAS Printers Limited, Wallop, Hampshire
Colour separations by Schwitter Limited, Zurich
Printed in Holland by Smeets, Weert

CONTENTS

BEFORE THE AEROPLANE

Gods and supermen

Why did men long to fly like the birds as far back in time as we can probe by studying the relics they left behind? Why, indeed, are so many of those relics concerned with flying, even when the different peoples who produced them were half a world apart, separated by great oceans or land masses that nobody ever crossed?

We can only guess at the answers to such questions. Some of our remote ancestors may have felt frustrated when they saw how effortlessly birds could wheel and soar through the skies while they – with infinitely greater powers of body and mind – could only move laboriously over land. A few, oppressed by their fellow men, captive or far from home, envied birds the freedom of flight. Typical was the psalmist, David, who cried: 'Oh that I had wings like a dove! for then would I fly away, and be at rest.'

As flying held the key to so much happiness, beyond the reach of enemies or the burden of worldly cares, it was natural that men should credit their primitive gods with the ability to fly. Every civilization from ancient China to Egypt, from Greece and Imperial Rome to Central and South America, had its winged gods.

Often, the wings were symbolic, signifying a belief that gods could be anywhere and everywhere, and know anything that happened, at any moment in time. The gentleness of a downy white wing could symbolize loving care; black leathery wings could represent the dark, overshadowing veil of death – for not all winged beings were envisaged as being sympathetic to humans.

So, from the gods and their heavenly companions, like Eros, Mercury and Hermes, the implied ability to fly was passed on first to lesser supernatural beings and then to highly-favoured mortals. It was stretching the imagination too far to credit these persons with the gift of natural wings. So, the Chinese Emperor Shun is said to have travelled by flying chariot in about 2200 BC, while Alexander the Great was envisaged as sitting in a cage hoisted by mythical winged griffins eighteen centuries later.

Throughout the
ancient world,
every nation had
its winged gods
and supermen.
These examples
are from Persia,
Rome, Greece,
Mexico and Egypt.

The bird-men

The first real progress towards human flight came with the downgrading of flying from the prerogative of gods to an attribute of legendary men and supermen. Wings sprouting naturally from shoulders or feet were clearly not credible; even the assistance of griffins and flying carpets stretched belief to the limit in the centuries of superstition and necromancy. So began the saga of the 'bird-men'.

Best-known of the early stories concerns Daedalus and his son Icarus. Imprisoned by Minos, king of Crete, for whom

he had built the Labyrinth, Daedalus made two sets of wings for an attempted escape. When the time came for the first flight test, he warned Icarus not to fly too high, lest the heat of the sun should melt the wax holding the feathers in place. Unfortunately, the young man found the experience of flying so exhilarating that he climbed higher and higher. Not until his wings disintegrated did he appreciate the wisdom of Daedalus' words – and then it was too late. Icarus fell into the sea and was drowned, while his father flapped on alone to the mainland of Italy.

Almost every country can produce similar stories as part of its folk-lore. The Scandinavians tell of Wayland the Smith, who soared to freedom on home-made wings after killing the sons of a king who had enslaved him. England has an even more notable 'bird-man' in the person of Bladud, ninth king of Britain and father of Shakespeare's King Lear. After discovering the mineral springs at Bath, by magical powers, he is said to have jumped from a high building in the city of Trinavantum, now London, wearing a pair of wings. This time his magic failed and he crashed to his death on the Temple of Apollo.

For hundreds of years after that, men continued jumping from towers and cliffs, vainly flapping their arms and plummeting, wide-eyed, to the ground. Oliver of Malmesbury, an eleventh-century British monk, is reputed to have flown more than a furlong (220 yards) before 'maiming all his limbs' in a heavy landing. In Scotland, John Damian was no more successful, but he knew why. He had, he said, made his wings from the feathers of chickens, which are ground birds, instead of from eagles' feathers.

'Bird-men' began with the legendary Daedalus and Icarus (*opposite top*) and King Bladud (*bottom left*), later included generations of tower-jumpers and the flying monk of Malmesbury

The pilot of Leonardo da Vinci's flying machine was supposed to use arms and legs to flap the wings, and head movements to control the tail

Leonardo's helicopter had a spiral wing to 'screw' itself up into the air

The genius of Leonardo

It was inevitable, perhaps, that the first fairly practical ideas on flying machines should have come from the great Leonardo da Vinci. Whilst acknowledging the beauty of his paintings, like the *Mona Lisa*, and the immense contributions he made to the early knowledge of anatomy, many modern critics imply that his other drawings depict only brilliant designs for machines and devices that were never built. They put them in the same category as the science-fiction writings of Jules Verne or H. G. Wells, and quote Leonardo's own reported last words on his death-bed: 'Was anything ever done?'

Aeronautically speaking, the answer to that question seems to be 'only on paper'. Leonardo produced one of the first drawings of a parachute, which he envisaged as a means of escape from a burning building. He devised a helicopter. The word comes from the Greek *helix* (a screw) and *pteron* (wing), and Leonardo's concept was just that – a spiral wing that would 'screw' itself up into the air.

What could he use to drive his wing? There were no mechanical power plants; not even any light metals from which to build his designs. Studying carefully the flight of birds, he switched his effort to designing aeroplanes that might enable a man to fly by utilizing his own strength to full advantage. These were no frail wings of feathers and wax fastened to the man's arms. A sturdy but light wooden framework supported pulleys and cables by which the man's arms and legs could be used to move wings modelled closely on those of real birds. One design even had a moving tail, controlled by movements of the would-be pilot's head, via a head-band and cable.

Leonardo's failure to build and test any of his aircraft probably resulted from his knowledge that they would never work. They were too heavy. No man could develop sufficient power to make them fly by flapping the delicate, beautiful wings. Just once, he came near to a practicable shape when he made the inboard part of each wing rigid and suggested flapping only the tips.

Such a machine might just have succeeded in flying as a glider, especially if the tips had not been flapped too enthusiastically. But it was another Italian, Giovanni Borelli, who had the last word, in 1680. In his book *De Motu Animalium*, he said that no man had the strength to fly by the unaided power of his own muscles. A few 'bird-men' still tried, and died.

Much later, in the early nineteenth century, a Swiss watchmaker named Jacob Degen claimed that he had flown with his flapping wings, but omitted to draw attention to the balloon that had raised him from the ground.

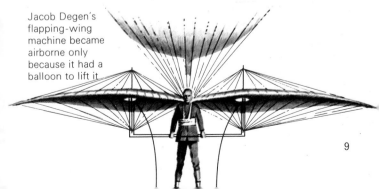

Jacob Degen's flapping-wing machine became airborne only because it had a balloon to lift it

Dew-drops and ethereal air

The rather smug, and portly, Chinese gentlemen shown below never really floated among the clouds in their paddle-wheel 'flying chariot'. In fact, it is no more than a representation of the vehicle said to have been used by the mythical inhabitants of the Kingdom of Ki Kouang. Kai Kawus, King of Persia nearly 3,500 years ago and builder of the Tower of Babylon, did not share the secret of such chariots. He is said to have become airborne by harnessing four geese to a mobile throne, and drawings produced many centuries ago imply that he not only travelled in considerable comfort but also took the precaution of arming himself against attack by other aerial voyagers. In fact, the bow and arrows clutched in his hands would certainly make him the world's first military pilot had his exploits been anything but flights of fancy.

Stories of the achievements of Kai Kawus – and Alexander the Great, who is said to have used griffins rather than geese as his 'power plant' – were taken very seriously by the people of bygone ages. On the other hand, the methods of flying illustrated opposite were intended to be regarded as mere figments of the imagination. They date from the sixteenth and seventeenth centuries when writers like Bishop

Godwin and Cyrano de Bergerac began telling tales about men on the Moon and journeys in fanciful spaceships.

Cyrano's favourite motive power was nothing more elaborate than the dew which can be seen on grass in the early mornings. Noting that this dew rose when the sun's rays fell upon it, he suggested simply putting a quantity of dew in the bottom of a spherical glass craft, which would then surely take off as soon as the sun rose. Ignoring the question of what happened if the sun hid itself behind a cloud when the craft had risen to a few hundred or few thousand feet, Cyrano described how his manned dew-drop was used for a journey to the Moon.

For commuting, he considered it necessary only to wear bottles of dew, or smoke, tied in strings around the waist and shoulders like life-belts. He was getting closer to a real lifting-force, but not as near as the thirteenth-century scientist monk, Roger Bacon, who is said to have conceived 'a large hollow globe ... filled with ethereal air, which would float on the atmosphere as a ship floats on water'.

Science-fiction writer Cyrano de Bergerac chose dew as the lift-force for his spacecraft and bottled smoke for more personal flight

Lifting agents for early flights of fancy ranged from aerial paddle wheels to the geese harnessed to Kai Kawus' throne (*opposite left and right*)

(*Above*) detail from fifteenth-century painting of *Madonna and Child*. (*Right*) de Lana's flying boat of 1670.

de Lana and Gusmão

It was all very well to suggest the use of 'ethereal air' as a lifting agent; where did a would-be flyer find it?

We know today that no such substance exists. We know, too, that for countless centuries people held the key to *real* flying in their hands every time they flew a kite. As an alternative, they might have tried to produce a large, man-carrying version of the familiar Chinese 'flying-top' toy, as shown in the hands of the baby Jesus in the remarkable painting on which the illustration at the top of this page is based. This might have led to a form of helicopter hundreds of years ago.

Three basic discoveries or inventions were needed to make flying possible. The first was a lifting force to raise man and machine off the ground; the second was a means of propulsion when the craft became airborne; the third was a method of controlling the direction of flight.

At first the pioneers concentrated mainly on 'lift'. This was understandable. They believed, naively, that once in the air they would be able to navigate by using sails and oars – as with a boat – or some kind of flapping wings.

Monks and priests tended to dominate the story of flying from the eleventh to the eighteenth centuries, because the monasteries of western Europe were the centres of learning and advanced scientific thought at that period. So it is no surprise that the first known design for a complete lighter-than-air craft, intended to 'float on the atmosphere', should have come from a Jesuit priest, Francesco de Lana-Terzi.

Twenty years after Otto von Guericke's invention of the air-pump in 1650, de Lana suggested that if all the air could be pumped out of a sphere, made of thin copper, it would become lighter than air. If four such spheres were then attached to a boat it would fly. A good idea – spoiled only by the fact that if the spheres were made of sufficiently-thin metal they collapsed when the air-pump got to work. If, on the other hand, they were made strong enough to prevent collapse, they were too weighty ever to become lighter-than-air.

Another Jesuit, Laurenço de Gusmão, came nearer to success in Lisbon, in 1709. His design for an aircraft, named the *Passarola* (Great Bird), may look crude, even humorous; but there is reason to believe that it flew in the form of a model glider – the world's first – and the strange 'sail' over the top hints at yet another probable 'first' that can be credited to Gusmão.

Gusmão's model Passarola may have flown as a glider

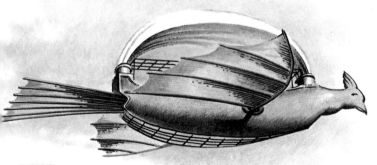

Hot air and hydrogen

In the same year that Gusmão built the *Passarola*, he demonstrated a mysterious model before the King of Portugal. Contemporary reports describe it as having a boat-like hull, with a *Passarola*-type parachute sail over the top. When a small fire was lit beneath the model, the sail filled out and the craft rose into the air. On the way down, it set fire to the King's curtains, but His Majesty must have been a kindly person for he was 'good enough not to take it ill'.

This event is both interesting and important, because it happened seventy-four years before two Frenchmen named Joseph and Etienne Montgolfier are said to have invented the hot-air balloon. The idea came to them one day while they sat by the fire at home, watching little pieces of paper being wafted upwards with the smoke. They knew nothing about heated air becoming rarefied and rising above cold air, but felt sure that if they trapped in a bag some of the mysterious 'gas' produced by the burning fire it would be powerful enough to lift objects off the ground.

They tested their theory first with a small silken bag, held open end downward over a fire. When released it rose to the ceiling. On 5 June 1783, they sent up a huge linen balloon, 38 feet in diameter, in front of the astonished inhabitants of their home town of Annonay. After that there was no holding them back. On 19 September an even bigger *Montgolfière* balloon was released in Paris, before the King and Queen of France, carrying in a basket the first living

(*Below*) Joseph Montgolfier and the balloon which carried a sheep, a cock and a duck on 19 September 1783

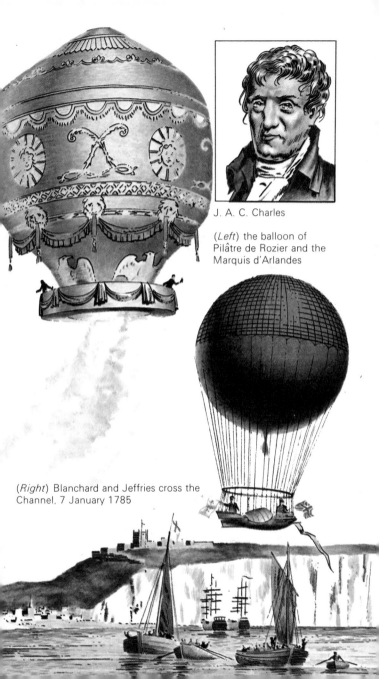

J. A. C. Charles

(*Left*) the balloon of Pilâtre de Rozier and the Marquis d'Arlandes

(*Right*) Blanchard and Jeffries cross the Channel, 7 January 1785

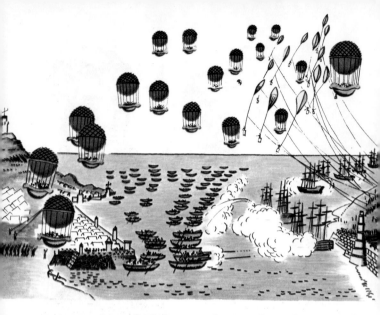

creatures ever to fly by man-made aircraft – a sheep, a cock and a duck.

It only remained for a twenty-nine-year-old physician named Jean-François Pilâtre de Rozier to ascend to a height of 85 feet under a huge *Montgolfière* on 15 October. Human flight was at last possible. To demonstrate his confidence in the new art, de Rozier travelled $5\frac{1}{2}$ miles across Paris on 21 November 1783, accompanied by the Marquis d'Arlandes. The Marquis was no mere passenger. Under the command of the 'pilot' he had to stoke the fire that produced 'Montgolfier gas' to keep the balloon airborne, and when the fire set light periodically to the balloon he had to dash over with a wet sponge to avert disaster.

Meanwhile, another Frenchman named Professor J. A. C. Charles had found an alternative to 'Montgolfier gas' which was, in reality, only hot air.

Back in 1766, an English chemist named Henry Cavendish had discovered a gas which he called 'inflammable air' and which we now know as hydrogen. One of its qualities was that it weighed less than one-tenth as much as an equivalent volume of air, and Charles realized that it would make an

Great flights of the nineteenth century included the scientific ascent of Coxwell and Glaisher to around 25,000 feet and Charles Green's 480-mile journey to Weilburg

(*Left*) Napoleon considered using balloons to invade England

ideal lifting agent. So he designed a balloon that could be filled with hydrogen and had it built by two brothers named Robert who had perfected a method of rubberizing silk to make it leak-proof.

Launched on 27 August 1783, the first unmanned Charles balloon ended its flight in the village of Gonesse. Believing their visitor from the sky to be a device of the devil, the villagers attacked it with pitchforks, then tied the wheezing, deflating 'carcase' to the tail of a horse, which raced over the countryside, tearing it to pieces.

Undeterred, Charles built a full-size version in which he made a two-hour flight with Marie-Noel Robert on 1 December that year. Seldom has any prototype been so advanced, for it embodied almost all the main features of a modern balloon.

On 7 January 1785, Jean-Pierre Blanchard of France and Dr John Jeffries, an American, made the first aerial crossing of the English Channel in a *Charlière*. Captive versions were used for battlefield reconnaissance by French armies from 1793, and by the mid-nineteenth century balloons were being used for everything from spectacular public displays to the first high-altitude research in Earth's atmosphere.

Birth of the airship

The balloon provided little but lift. Once airborne, its occupants could travel only as fast as the wind carried them, in the direction it happened to be blowing. A journey was strictly one-way; after landing it was usual to deflate the envelope, pack it into a basket and return home by train. This was very different from the sort of flying of which man had dreamed for centuries – the unrestricted, fast, wheeling and soaring which even the most common birds could enjoy.

No-one could deny that the balloon represented a step in the right direction. It got men into the air and it had many practical uses. This was made abundantly

(*Top left*) balloon leaving Paris in 1870. (*Left*) Jullien's *Précurseur*, first model airship. (*Below left*) Cayley's design for an airship. (*Below*) Henri Giffard's airship, 1852.

clear when the Prussian army laid siege to Paris in 1870. Cut off from the rest of France, the people of the capital organized the first-ever air service, using balloons for one-way journeys that carried out of their city more than 100 passengers, including the Prime Minister, nine tons of mail and over 400 carrier pigeons, of which 57 found their way back. As a pioneer air lift it was superb, but only the Prussians were thankful that return journeys were impossible.

An English baronet named Sir George Cayley – who was to become recognized as one of the greatest men in flying history – had suggested as early as 1837 that the 'gas-bags' of balloons should be made streamlined, and designed a craft which also had steam-driven propellers for propulsion and steering. It was never built, but in 1850 Pierre Jullien of France constructed a clockwork-powered model of similar shape, which he named the *Précurseur*. It inspired another Frenchman, Henri Giffard, to build a full-size version, powered by a 3-hp steam-engine, which is honoured today as the first airship. It was, in fact, only partially successful, for although Giffard flew it from Paris to Trappes at a speed of 6 mph in 1852, its controllability was marginal.

Not until Renard and Krebs built and flew the electrically-powered *La France*, in 1884, did the airship progress to the stage of being sufficiently controllable to be steered back to the place from which it began its journey. *La France* was big, with a length of 165 feet, and was quite fast by the standards of the time, with a top speed of $14\frac{1}{2}$ mph. What it lacked mostly was an efficient power plant. Even this was about to be remedied, for the lightweight four-stroke internal combustion engine had already been invented in Germany.

First completely controllable airship was *La France*, built by Renard and Krebs in 1884

Santos-Dumont mated the airship and the petrol-engine (*above*) ; Zeppelin perfected the big rigid airship (*below*).

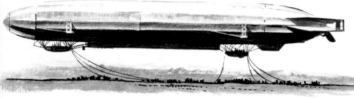

Non-rigid to rigid

It was a dapper little Brazilian named Alberto Santos-Dumont, living in Paris, who first mated successfully the airship and the petrol engine. The date was 1898, but the exploit which made him the idol of French society occurred three years later when he captured a 125,000-franc prize by making the first flight from Saint-Cloud, round the Eiffel Tower and back in under thirty minutes.

Santos-Dumont's airships were small 'non-rigids' − i.e. their envelopes kept their shape because of the gas contained inside. Very different were the giant aircraft built in Germany by Count (Graf) Ferdinand von Zeppelin. No mere elongated balloons, they were real 'flying ships', with separate gas containers retained inside a rigid metal framework, covered with fabric.

Construction of the first of them was started in 1898 and completed two years later. On its second flight, it buckled so severely that he could not afford to repair it. That might have been the end of the Zeppelin story had he not been permitted to organize a State lottery to raise the money for a replacement. This airship, too, was wrecked in a gale. More lotteries gave birth to more airships, and between 1910 and 1914 the Zeppelin company operated the first-ever regular passenger air services between Lake Constance, Berlin and other cities in Germany. The twenty-seat *Deutschland* and her successors carried a total of 35,000 passengers before the outbreak of the First World War brought their operations to a close.

For four years, the Zeppelins shadowed allied naval forces at sea and pioneered strategic bombardment, until the increasing capabilities of fighter-planes made them too vulnerable. With the return of peace, the Zeppelin company raised money to build the 772-foot-long *Graf Zeppelin*, which made the first of more than 100 transatlantic flights, to New York, in 1928, carrying a crew of forty and twenty passengers. Later came the larger and even more luxurious *Hindenburg;* but on 6 May 1937, whilst this great airship was being moored in America, it burst into flames. Thirty-five people lost their lives, and with them died the big rigid airship.

No other aircraft have matched the interior spaciousness and comfort of the *Graf Zeppelin* and *Hindenburg*

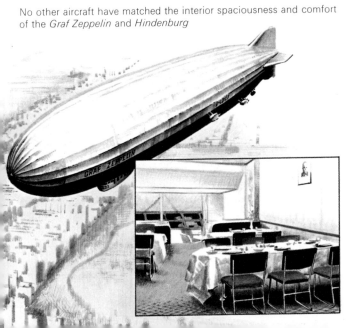

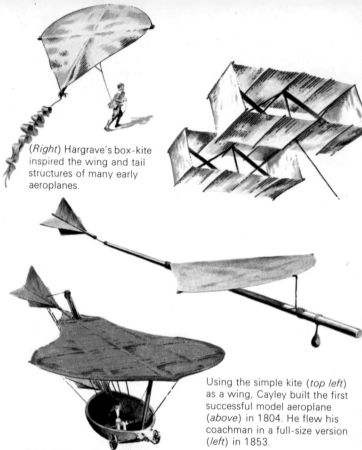

(*Right*) Hargrave's box-kite inspired the wing and tail structures of many early aeroplanes.

Using the simple kite (*top left*) as a wing, Cayley built the first successful model aeroplane (*above*) in 1804. He flew his coachman in a full-size version (*left*) in 1853.

MOSTLY FANTASY

Kites and Cayley

What was the alternative to lighter-than-air flying, which had cost so many lives?

The Chinese had been flying kites from time immemorial. No mere toys, some were large enough to carry a man into the air. Marco Polo described how man-lifters were sometimes sent aloft before a planned sea voyage. If they flew steadily, it was a good omen; if they crashed, killing or injuring their passengers, it was regarded as a warning that the sailing should be postponed.

It is doubtful if anybody thought much about why kites flew; all that mattered was that they did – usually. But towards the end of the eighteenth century, Sir George Cayley – who, as we have already seen, became interested later in airships – began to think about the forces that would act upon a heavier-than-air aeroplane in flight.

One of the treasures of London's Science Museum is a small silver disc, engraved by Cayley with his first design for an aeroplane, based on the theories he had evolved. The sail-wing is reminiscent of Gusmão's *Passarola*; but, for the first time lift was divorced from thrust. The sail-wing was envisaged as providing lift from the air, as does a kite, and Cayley suggested the use of large paddles for propulsion. Recognizing the need for control, as well as lift and thrust, he added tail surfaces.

From this disc of 1799, it was but a short step to his five-feet-long model glider of 1804. First heavier-than-air craft capable of proper flight, it consisted of a kite-wing mounted on a pole, and movable cruciform tail surfaces. Cayley improved the lift produced by the kite by raising its leading-edge – giving it what we now call an angle of incidence. He 'balanced' the aircraft, or put its centre of gravity in the right place, by means of a moving weight, suspended beneath the pole.

Cayley wrote in his notebook: 'It was very pretty to see it sail down a steep hill, and it gave the idea that a larger instrument would be a better and safer conveyance down the Alps than even the surefooted mule'.

In 1853, when he was eighty years old, Cayley built his 'larger instrument' and used it to fly his elderly coachman across a valley on his estate. The man resigned afterwards, protesting that he was hired to drive not fly.

Henson's Aerial Steam Carriage, an inspired design of 1842, owed much to Cayley

23

The first powered hops

In 1846, four years after William Samuel Henson designed his Aerial Steam Carriage, he referred to Cayley as 'the father of aerial navigation'. This was justified. Cayley had not only devised the layout of the aeroplane as we know it today and conducted the first man-carrying heavier-than-air flight; he had also laid down such basic principles as the value of streamlining, the need for wings to be cambered, the way to produce a sturdy wing structure by adopting a wire-braced biplane or triplane layout (with two or three wings one above the other), and the use of dihedral (with the wings forming a flattened V in head-on view) to improve stability.

In the Aerial Steam Carriage, Henson used the teachings of Cayley to inspire the design of an aeroplane that looks practical even now. To finance its construction, he formed the Aerial Transit Company and issued pictures showing the Steam Carriage in flight over London, the Pyramids and other places, to attract investors. The only reaction was ridicule.

Henson made a 20-feet-span model of his design, which can still be seen in the Science Museum, London. It failed to fly because there was no engine to make this possible.

In France, a naval officer named Félix du Temple designed a monoplane with the propeller in the best possible place, at the front, and even suggested building it of lightweight aluminium, then a very new material. A model of it, built and tested in 1857, first with a clockwork motor and later steam-powered, became the first heavier-than-air craft to take off and fly under its own power. Seventeen years later, a full-size version made a short hop-flight at Brest, piloted by a young sailor, but only after gathering speed down a slope.

The big steam-powered monoplane of Alexander Mozhaisky made a similar 'assisted flight' in Russia in 1884. Ten years later in England, Sir Hiram Maxim, inventor of the machine-gun, came even nearer to success. He did not intend to fly and anchored his huge biplane to a 'railway track' to keep it down; but it developed so much lift that it broke free of the guard rails and flew briefly.

Before then, on 9 October 1890, Clément Ader of France had persuaded his bat-winged Eole to hop 150 feet, so becoming the first full-size aeroplane to lift itself from level ground.

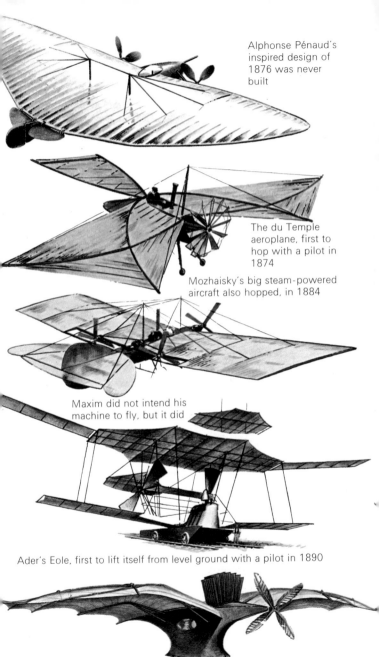

Alphonse Pénaud's inspired design of 1876 was never built

The du Temple aeroplane, first to hop with a pilot in 1874

Mozhaisky's big steam-powered aircraft also hopped, in 1884

Maxim did not intend his machine to fly, but it did

Ader's Eole, first to lift itself from level ground with a pilot in 1890

Gliders point the way

Ader's Eole and later Avion III were less practical designs than those of Henson, du Temple and Mozhaisky. What they indicated, more than anything else, was that nobody would fly properly until the theories of Cayley were matched with a lightweight, efficient power plant.

One of the first to appreciate this was Otto Lilienthal, in Germany. Recognizing that powered flight was still not practicable, he began experimenting with a series of gliders. Some survive in museums, and they are among the most beautiful aircraft in history, with bird-like wings of peeled willow wands covered with waxed linen cloth.

In six years Lilienthal made thousands of successful gliding flights in these aircraft, proving conclusively that heavier-than-air flight was possible. He made an artificial hill from which to take off and covered distances of more

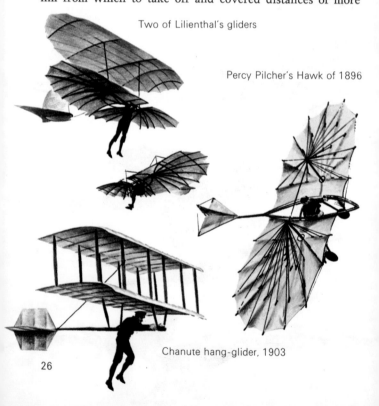

Two of Lilienthal's gliders

Percy Pilcher's Hawk of 1896

Chanute hang-glider, 1903

26

than 750 feet, up to 75 feet above the ground. He controlled his gliders in flight by hanging beneath them and swinging his body to and fro, and sideways, to steer and stabilize them.

By 1896, he had progressed to the stage where he was about to fit a small engine to one of the gliders; but on 9 August that year, at the age of only forty-eight, he paid the price for ignoring more efficient methods of control. Losing his balance in the air, he crashed and was fatally injured. His last words were 'Sacrifices must be made', and they have been true of every phase of flying, in the atmosphere and out into space.

Among the disciples of Lilienthal were Percy Pilcher in England and Octave Chanute in America, both of whom flew successfully a number of 'hang-gliders'. Pilcher, too was killed just as he was about to fit a 4-hp oil engine to one of his designs. Chanute is remembered not only for his sturdy biplane gliders but because he collected together in a book, entitled *Progress in Flying Machines*, information on all significant experiments made by the 1890s.

Chanute's writings, and the flights of Lilienthal, so inspired two young bicycle-makers of Dayton, Ohio, named Wilbur and Orville Wright, that they became determined to fly. Beginning with a pilotless glider, flown like a kite at the end of cables, they soon evolved a workable control system. Finding all recorded data on wing shapes unreliable, they developed their own wing sections with the help of a home-made wind-tunnel. By 1902, they were able to fly so well that the time had clearly come to try fitting an engine.

Wright No. 1 glider flying as a kite, 1900

Wright No. 2 glider, 1901

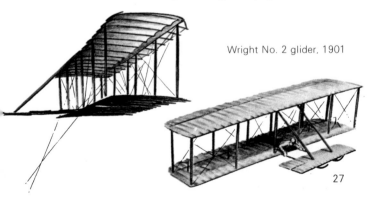

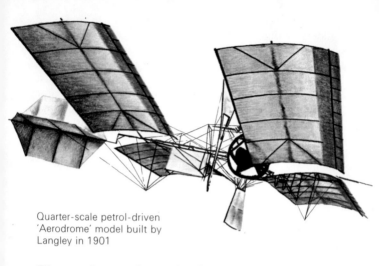

Quarter-scale petrol-driven
'Aerodrome' model built by
Langley in 1901

The professor from the Smithsonian

Before the Wright brothers could complete their powered
'Flyer', a formidable competitor took the stage briefly.

Samuel Pierpont Langley, Secretary of the famous Smithsonian Institution, first showed his talents in 1896, when he
built a 16-feet-span tandem-winged model aeroplane powered
by a 2-hp steam-engine, which succeeded in flying well over
half a mile. Impressed, the US War Department offered him
$50,000 towards the cost of building and testing a full-size
version. Bearing in mind the general disbelief in a future for
heavier-than-air flying at that time, this is a tribute not only
to the far-sightedness of the government but to the reputation of Langley and the Smithsonian at a time when would-be
flyers were regarded generally as starry-eyed fanatics.

If careful, professional design had been sufficient to ensure
success, we should today honour Langley as the man who
created the first practical powered aeroplane. His 'Aero-

The first test of the full-size 'Aerodrome', 7 October 1903

drome', as he called it, was similar in layout to the 1896 model, and was fitted with an ingenious 52-hp five-cylinder petrol-engine which set the pattern for all later 'radials', with cylinders arranged star-shape and a central crankshaft. The engine was produced by Charles Manly, who also volunteered as test pilot.

Twice the 'Aerodrome' was launched from a specially-equipped houseboat on the Potomac river – the idea being, presumably, that a subsequent touch-down on water would be 'softer' or less hazardous than a landing ashore. On the first occasion, on 7 October 1903, it hit a post on the catapult launching gear and plunged into the river. By 8 December, Manly and the 'Aerodrome' were ready for a second attempt. Again the tail unit fouled the launch gear, collapsed, and the aircraft fell into the water.

With a typical, uninformed attempt at humour, one of the journalists who had witnessed the trials wrote: 'If Professor Langley had only thought to launch his air-ship bottom up, it would have gone into the air instead of down into the water'. Dreadfully disappointed, the sixty-nine-year-old pioneer gave up his experiments. Many years afterwards, in 1914, after making changes to the structure, a later great American designer named Glenn Curtiss succeeded in flying the 'Aerodrome' successfully as a seaplane; but by then the Wrights had established their place in history as the creators of the aeroplane.

The second unsuccessful launch, 8 December 1903

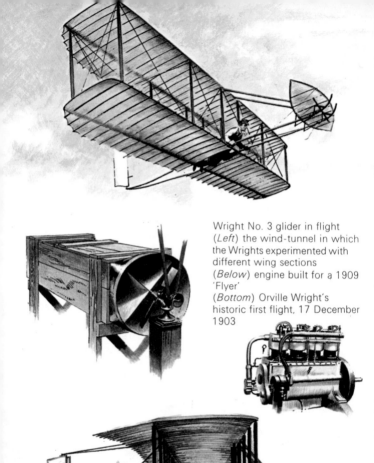

Wright No. 3 glider in flight
(*Left*) the wind-tunnel in which the Wrights experimented with different wing sections
(*Below*) engine built for a 1909 'Flyer'
(*Bottom*) Orville Wright's historic first flight, 17 December 1903

Success at Kitty Hawk

The glider in which Orville and Wilbur Wright made hundreds of successful flights in 1902–3 was quite large, with a wing span of 32 ft 1 in. Its control system comprised a forward elevator for climb and descent, and a series of cables linked to a cradle which could be rocked by the pilot's hip movements to warp, or twist, the wing-tips.

Unfortunately, the drag of the warped tips, combined with the effect of fixed tail-fins, often turned the glider in the opposite direction to that intended, and even caused it to spin into the ground. This was cured by replacing the fins with a single movable rudder, linked to the warp-cradle so that it always turned towards the warping direction. The rudder then not only offset the drag but enabled the glider to make smooth, banked turns.

It was this glider design that the Wrights scaled up to produce their 1903 'Flyer'. Spanning 40 ft 4 in, this had warping wings, a biplane front elevator and twin rudders. The 12-hp four-cylinder engine, built by the brothers, drove two pusher propellers through chains. Another link with their bicycle business was found in the undercarriage, as the aircraft's skids were designed to run along a launching-rail on a small trolley fitted with rollers made from cycle wheel-hubs. The pilot lay on the lower wing, with his hips in the wing warp-cradle.

When the Wrights took their 'Flyer' to Kill Devil sand hills, near Kitty Hawk, North Carolina, in October 1903, they spent the early part of that month brushing up their flying on the No. 3 glider. Only when they learned of the first unsuccessful Langley test did they realize that they might be beaten into the air.

Before all was ready for the first test of the 'Flyer', Langley had tried again, and failed. The Wrights were no more fortunate on 14 December when Wilbur moved the elevator too soon, half-lifted off the launch-track and slammed back into the ground.

It was Orville's turn next, and at 10.35 am on 17 December 1903, he made the first 12-second flight in the 'Flyer', covering 120 feet.

At long last, man had grown wings.

GROWING WINGS

The stick-and-string era

In all, the Wrights made four flights on 17 December 1903. The final one, by Wilbur, lasted 59 seconds and covered 852 feet. For an expenditure of under $1,000, spread over four years, the brothers had logged a total of 97 seconds in powered flight. This represented the entire active life of 'Flyer No. 1', which was overturned by a gust of wind and wrecked soon after Wilbur's landing; but it was enough.

Its achievement drew no sensational headlines in the world's press, because editors had learned to be sceptical about claims to have flown. The Wrights tried to offset this by inviting newspaper representatives to witness the first take-off of their improved 'Flyer No. 2' at the Huffman Prairie, eight miles from their home in Dayton, in May 1904.

Santos-Dumont 14 *bis*

The first Demoiselle, 1907

Farman's improved version of the Voisin biplane

It would have been better had they first tested the aircraft. With a faulty engine, it simply ran off the end of its launching rail and came to a stop.

After that the press showed little interest, even when passers-by reported having seen the Wrights in the air, flying properly for long periods. Orville and Wilbur no longer cared. Confident that no-one else would build a worthwhile aeroplane within five years, they gave up flying and simply waited for someone to appreciate the value of their aeroplane.

Nineteen days after 'Flyer No. 3' had covered twenty-four miles non-stop in a flight on 5 October 1905, the US Board of Ordnance and Fortification had told them that the Army would not be interested 'until a machine is produced which by actual operation is shown to be able to produce horizontal flight and to carry an operator'.

In one respect, the lack of knowledge of what the Wrights had achieved, and of the form of their aircraft, proved a godsend. Its 'back-to-front' forward elevator design was a dead end, and its absence of a wheeled undercarriage, which tied it to a launch track at its base, was a further drawback.

Left largely to their own resources, pioneers in Europe were not to build better aeroplanes than the Wrights for another three years; but in their ignorance they stumbled eventually on a better basic layout.

Santos-Dumont, the airship builder, was first to fly in Europe, on 23 October 1906, in his ungainly tail-first 14 *bis*. Its wings and tail were box-kites, of the type perfected by Lawrence Hargrave in Australia in 1893. By 1908, he had progressed to the tiny Demoiselle, intended as the first 'build-it-yourself' aeroplane.

On 13 January 1908, Henry Farman won a 50,000-franc

Roe's first Triplane, 1909

Curtiss Gold Bug, a 1909
development of the June Bug

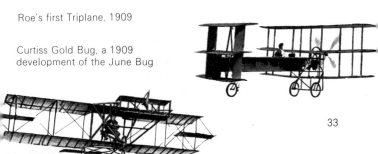

Rheims 1909. An Antoinette leads, successively, a Wright biplane, Curtiss Golden Flyer, Voisin, Blériot XI and another Voisin.

prize by completing the first circular flight of more than one kilometre on a front-elevator Voisin 'box-kite', and went on to evolve improved versions of the design. His American counterpart was Glenn Curtiss, who won a trophy presented by the *Scientific American* magazine by flying nearly a mile in his 40-hp June Bug on 4 July 1908.

Curtiss was one of five members of the Aerial Experiment Association formed by the Canadian inventor Alexander Graham Bell. He was to find even greater fame later, as the true pioneer of water-borne aircraft. Meanwhile, one of his colleagues, J. A. D. McCurdy, made the first official powered flight by a British subject in the British Empire, in his Silver Dart biplane, on 23 February 1909, from the ice-covered Baddeck Bay in Canada.

McCurdy may not really have been first. There is strong evidence to suggest that Horatio Phillips covered some 500 feet in a tandem-winged aeroplane in England in the summer

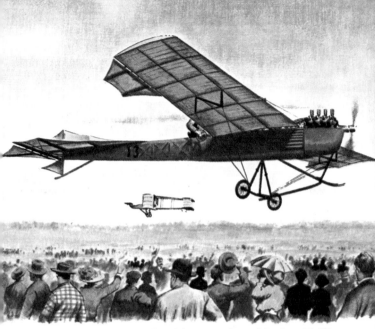

of 1907. We shall never be able to confirm this, but Phillips deserves a place in our story. His aeroplanes looked like Venetian blinds on wheels; but each 'slat' was a beautifully-made, narrow-chord, cambered wing.

British would-be aviators had to contend with both public ridicule and official hindrance. Of none was this more true than A. V. (later Sir Alliott Verdon-) Roe. After winning a £75 prize at the *Daily Mail*'s model aeroplane competition in 1907, he built a full-size version of his aeroplane and succeeded in hopping briefly along Brooklands motor racing track on 8 June 1908.

Unwanted there, he moved to Lea Marshes and built a tiny triplane with a 9-hp engine, covering it with brown paper to save weight and money. In this machine, on 13 July 1909, he became the first Briton to fly an all-British aeroplane in the UK, and went on to become the greatest of all the country's pioneers.

In contrast, aviators were regarded as heroes in France, and more than a quarter of a million people flocked to Rheims in

August 1909, to watch most of the world's great pilots compete in the first-ever air races. It was an unforgettable experience. Records were broken every day, and nobody was killed, although several aeroplanes were wrecked.

No more moats

One of the pilots who crashed at Rheims in 1909 was Louis Blériot. This was no surprise, as he was far better known at first for his hair's-breadth escapes from death than for his successes. Yet one of those successes was to put him among the handful of pilots whose names would be remembered through the years by people with little general interest in aviation.

The *Daily Mail* prize that A. V. Roe had won in 1907 was the first of a whole series offered by that newspaper in the previous year. For example, £1,000 was put aside for the first pilot to fly an aeroplane over the English Channel, and £10,000 for anyone able to link the paper's two publishing centres, London and Manchester, by air within twenty-four hours. One of the *Daily Mail*'s rivals proclaimed sarcastically that he would happily give £10 million to anyone able to perform such a feat.

How soon did that rival editor begin to regret his haste? Was it in October 1908 when Louis Blériot, as a rehearsal for his planned cross-Channel attempt, made a $25\frac{1}{2}$-mile cross-country flight? In doing so, he had made three intermediate landings. This time there could be no stops *en route*.

Blériot was not alone in his willingness to face the twenty-two miles of water which had long been Britain's moat, cutting her off from Europe's petty squabbles and wars. Before he was ready to start, Hubert Latham set out in a graceful Antoinette monoplane on 19 July. There was no cooler character in early aviation. Latham had taken up

The Blériot XI monoplane, first to fly the Channel

flying because doctors had said he had only a year to live and he considered that he might just as well live dangerously. When he and his aircraft were hauled out of the Channel, only two miles from France, he was sitting on top, smoking a cigarette, and had not even got his feet wet.

Before Latham could try again, Blériot took off from Baraques, near Calais, on 25 July 1909. Seldom more than a few feet above the water and with no compass, he lost his way in mid-Channel. Then his 24-hp Anzani engine began to overheat. Just in time, a providential shower cooled it and Blériot went on to land in a meadow behind Dover Castle. The *Daily Mail* £1,000 had been won, and the £10,000 prize was within reach thanks to a revolutionary new aero-engine designed in France – the 50-hp Gnome.

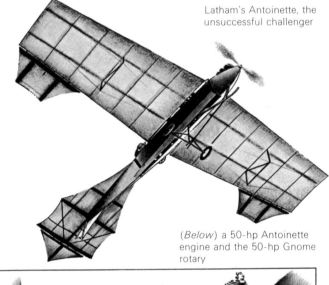

Latham's Antoinette, the unsuccessful challenger

(*Below*) a 50-hp Antoinette engine and the 50-hp Gnome rotary

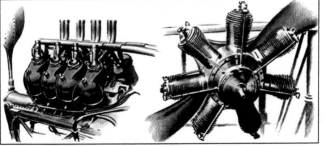

Building big

The Gnome engine was revolutionary in more ways than one. It was what is known as a rotary engine, which means that its crankshaft was bolted to the airframe and the cylinders rotated around it, taking the propeller with them. This improved cooling, at the expense of vast quantities of castor oil lubricant which was exhausted out of the cylinder ports.

Outstanding aeroplane of 1909 was the Voisin biplane. Of twenty-three machines which flew at Rheims, seven were Voisins; but it was Henry Farman's improved version, with a 50-hp Gnome engine, which succeeded it and became the great prize-winner of the era. In an aircraft of this type, Louis Paulhan won the £10,000 *Daily Mail* prize, by flying from London to Manchester in twelve hours in April 1910.

Gnomes were built eventually with two rows of cylinders, giving up to 160-hp; but reliability suffered and, initially, it seemed better to fit more engines rather than have a single more powerful engine. Among the first were the twin-engined Shorts of 1911, with engines fore and aft of the pilot. In Russia, Igor Sikorsky went even further by building *Le Grand,* the world's first four-engined aeroplane, in 1913 and following it with the even bigger *Ilia Mourometz*. This had an enclosed cabin and a promenade deck above the fuselage on which intrepid passengers could walk in flight.

Less imaginative, but equally significant for the future, was Claude Grahame-White's Charabanc, used for joy-riding at Hendon aerodrome. Spanning 62 ft 6 in and powered by a 120-hp engine, it carried nine passengers for nearly twenty minutes on 2 October 1913, leaving little doubt that fixed-wing airliners would one day replace the Zeppelin airships then used on the first airline services.

Gnome-engined Voisin flown at Rheims by 18-year-old Etienne Bunau-Varilla

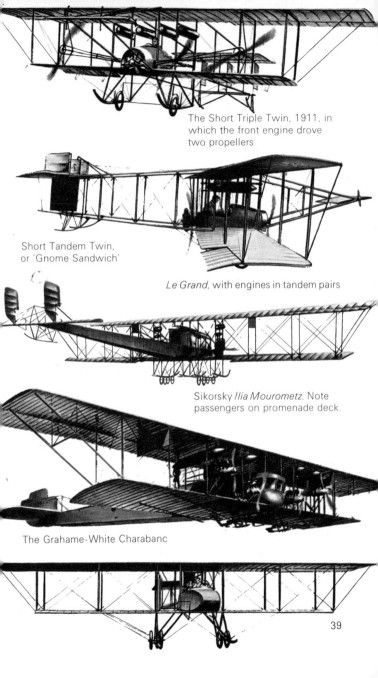

The Short Triple Twin, 1911, in
which the front engine drove
two propellers

Short Tandem Twin,
or 'Gnome Sandwich'

Le Grand, with engines in tandem pairs

Sikorsky *Ilia Mourometz*. Note
passengers on promenade deck.

The Grahame-White Charabanc

First on floats

With few proper aerodromes available in aviation's first decade, waterways seemed to offer the least cluttered areas for flying.

First to fly an aircraft with a float undercarriage was a Frenchman named Henri Fabre. His aircraft, which is preserved in the Musée de l'Air near Paris, looks very strange even by the standards of its time. It was a tail-first design, with no fuselage for the pilot to sit in. Instead, he perched atop an open rectangular framework, and the general appearance of the seaplane was not improved by the lattice girder spars that formed the basic structure of wings and tail. These were believed by Fabre to offer the best combination of high strength and low drag, as the airflow passed through them.

Fabre flew his seaplane for the first time at Martigues on 28 March 1910. It remained a 'one-off' freak; but by the autumn of that year, in America, Glenn Curtiss was already beginning to demonstrate the value of aeroplanes as reconnaissance aircraft for the navy. On 14 November, Eugene Ely flew a Curtiss biplane off a wooden staging above the foredeck of the cruiser *Birmingham*. Two months later, he landed a similar aeroplane on the USS *Pennsylvania*. So was born the concept of the aircraft carrier.

These Curtiss biplanes were not seaplanes, but carried flotation tubes to prevent their sinking if they alighted accidentally on the water. The next step was to replace the wheels with a pair of floats on which the aircraft could take off and land. This Curtiss did, flying the world's first really practical seaplane on 26 January 1911.

Soon he was regarded as the leading exponent of water flying. He progressed from seaplane to flying-boat by first fitting a machine with a single wide float, instead of two narrow ones, and then putting seats inside the float.

A variation of this idea, tried out on the Radley-England Waterplane, was to seat the crew and five passengers in two large separate floats. This was not the only unique feature of the aircraft, because it had three 50-hp Gnome engines, mounted one behind the other and driving a single pusher propeller through chains.

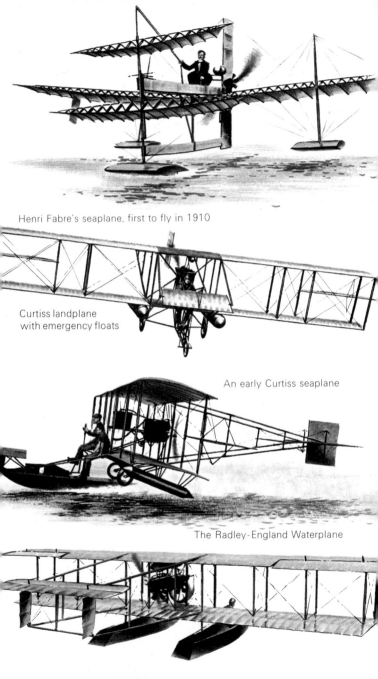

Henri Fabre's seaplane, first to fly in 1910

Curtiss landplane
with emergency floats

An early Curtiss seaplane

The Radley-England Waterplane

Putting aeroplanes to work

Blériot's achievement in crossing the Channel, and the steady improvement in reliability of aircraft and engines, gradually convinced the authorities and the public that aviation might, after all, have a future. With predictable ingenuity, designers soon began fitting guns and bomb-racks on aeroplanes. However, such experiments are outside the scope of this book, which is devoted to more peaceful uses of the aeroplane.

The first civilian task which was found for the aeroplane had its origins in the air mail balloon service out of Paris in 1870. The date was February 1911 and the occasion an exhibition at Allahabad, India. Captain (later Sir) Walter Windham happened to be there with a Humber biplane and a French pilot named Henri Pequet. At Windham's suggestion, the Indian Post Office cut a special cancelling stamp reading 'First Aerial Post, UP Exhibition, Allahabad 1911'. Pequet then flew bags of mail franked with this stamp from the

Blériot XI used on 1911 Coronation Mail service

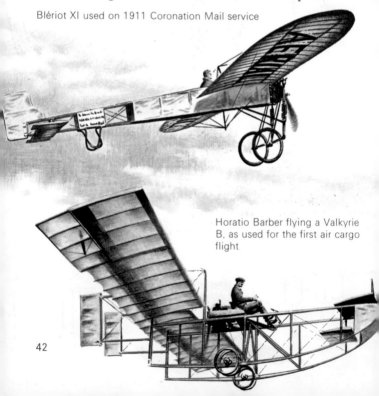

Horatio Barber flying a Valkyrie B, as used for the first air cargo flight

grounds of the exhibition, across the Jumna river, to the post-office at Naini.

The service was a great success, more than five thousand letters being carried in one day. So, to commemorate the coronation of King George V later that year, Windham suggested that a similar operation ought to be mounted between Hendon aerodrome and Windsor. The Postmaster General in London proved receptive to the idea, and during ten days in September 1911 about 100,000 specially-printed letters and cards were flown over the route by pilots from the Blériot and Grahame-White schools.

One pilot crashed on take-off and broke his legs. Nevertheless, it is difficult to agree with the verdict of an historian, many years later, that 'the experiment proved very little beyond the fact that aircraft were still hazardous vehicles in rough weather, and extremely unreliable in the matter of timing'. They were less than eight years old at the time!

The first air cargo was, reputedly, a box of Osram lamps carried from Shoreham to Hove, Sussex, by Horatio Barber in one of his Valkyrie tail-first monoplanes on 4 July 1911. Not for three more years was the first scheduled airline service opened with aeroplanes. Even then, the St Petersburg–Tampa Airboat Line carried only one passenger on each twenty-mile flight over Tampa Bay, Florida. It saved a long journey around the lake shore, but was maintained for only four months. Further progress had to await the end of a great war.

Benoist two-seat flying-boat of
the St Petersburg-Tampa
Airboat Line

The story of how pilots in tired D.H.4 mailplanes battled through dreadful weather to establish the US Air Mail service is one of the epics of aviation history

MAIL OR PASSENGERS?

Flying the US mail

At the start of the First World War, the sole task of the military aeroplane was to operate as a kind of aerial cavalry, keeping track of enemy troop movements for the benefit of friendly forces on the ground. By the time the war ended, aeroplanes had evolved into formidable strategic bombers and ground attack machines. They fought each other in the air, sought out submarines, launched torpedoes at enemy ships and flew from the decks of specially-built aircraft carriers at sea.

However, the machines of 1918 were no faster than the 131-mph S.E.4 biplane that had been built at the Royal Aircraft Factory, Farnborough, in 1914. None had a more roomy cabin than Sikorsky's pre-war *Ilia Mourometz*. Construction was still mainly of wood with fabric covering; cockpits were still open to the elements; radio had not progressed to the stage where it could be of much help to an airline. About the only positive gain was that aero-engines had become more powerful and far more reliable.

There were thousands of war-surplus aeroplanes, which could be bought at ridiculously low prices, and just as many war-trained pilots who hoped to continue flying. Disillusionment came swiftly. Pilots who tried to fly privately found the

Variants of the D.H.4 were legion. The Lowe, Willard and Fowler D.H.4B conversion (*above*) had two 150-hp engines and nose mail compartment. No. 299 (*below*) had special wings and underslung mail stowage.

cost crippling. Airlines soon discovered that the general public could not afford to fly, with fares on the short London-to-Paris route costing up to £21 one-way – equivalent to about £100 today.

The turning point in America came with the decision to operate experimental air mail services across the country. The Post Office began in 1918 with seventeen ex-Army D.H.4 biplane bombers, designed in Britain but built in the USA. It was a heartbreaking business trying to battle through the worst winter weather in these open-cockpit wood-and-fabric aeroplanes. Thirty of the first forty pilots recruited for the job died flying the mail, but eventually they created a day-and-night transcontinental service that cut the delivery time for mail by half. The government responded by spending more than half a million dollars on beacons to light the airways along which the mailplanes were to be followed one day by huge fleets of airliners.

The Vickers Vimy in which Alcock and Brown made the first non-stop crossing of the Atlantic by air

First across, via the Azores, was the US Navy Curtiss flying-boat NC-4

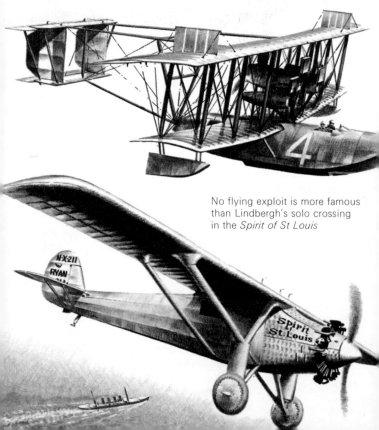

No flying exploit is more famous than Lindbergh's solo crossing in the *Spirit of St Louis*

Blazing the trail

Reference has been made already (pages 36–38) to the series of prizes offered by the *Daily Mail* newspaper to encourage feats of early flying. One major prize that remained unclaimed at the start of the First World War was a sum of £10,000 for the first crossing of the Atlantic Ocean.

An Englishman named John Porte was preparing for an attempted crossing in a twin-engined Curtiss flying-boat when the war started. Instead, he introduced improved versions of the aircraft into service with the Royal Naval Air Service and developed from them the family of Felixstowe 'F' flying-boats which performed so well over the waters around Britain.

As soon as the war was over, pilots began seeking ways of winning the *Daily Mail*'s £10,000, which was restricted to British crews in British aircraft. Before any of them was ready, on 16 May 1919, three big Curtiss flying-boats of the US Navy, each powered by four engines, took off from Newfoundland for a crossing in stages, via the Azores. Only one of them, the NC-4, completed the flight to Lisbon, in twenty-five hours' flying time, and then pressed on to England. The others were forced down at sea, fortunately without casualties.

A similar fate befell the first British crew, Harry Hawker and Lt-Cdr Mackenzie-Grieve, who left Newfoundland in the Sopwith Atlantic only two days after the Curtiss 'boats. The ship that picked them up had no radio, so they were mourned as dead for a week before reaching port.

Next to leave, at 4.15 pm on 14 June, were Captain John Alcock and Lt Arthur Whitten-Brown in a converted Vimy bomber. Lashed by storms, their radio out of action and almost deafened by a broken exhaust on one of the 360-hp Rolls-Royce Eagle engines, they droned on for hour after hour. At one stage, the big biplane became covered with ice and fell to within fifty feet of the wave-tops, almost on its back. Only Alcock's superb piloting saved them, and eventually, after $16\frac{1}{2}$ hours, they touched down in Ireland. What looked like a green meadow proved to be a bog, and the Vimy upended on to its nose; but Alcock and Brown were unhurt, one of the world's most frightening stretches of ocean had been flown non-stop and the trail had been blazed for an important air route of the future.

Other great flights followed, as the whole world was traversed by air. A Vimy piloted by Australians Ross and Keith Smith, with two mechanics, hopped doggedly over the 11,294 miles between Hounslow and Port Darwin in twenty-eight days in November-December 1919. Like Alcock and Brown, the pilots were knighted for their achievement.

Shortly afterwards, on 4 February 1920, Lt-Col Pierre van Ryneveld and Major Quintin Brand took off from Brooklands in another Vimy, bound for Cape Town. They wrecked the aircraft in a forced landing in North Africa, carried on to Bulawayo in a replacement Vimy, wrecked this one while taking off for Pretoria and reached Cape Town on 20 March in a D.H.9.

Two Portuguese naval Captains, Gago Coutinho and Sacadura Cabral, had similar experiences during their first crossing of the South Atlantic in 1922, completing the trip in three different Fairey floatplanes. Very different was the

First over the North Pole was Byrd's tri-motor Fokker

Kingsford Smith's *Southern Cross*

Byrd's Ford Tri-motor, first over the South Pole

$33\frac{1}{2}$-hour one-man one-plane New York-to-Paris non-stop flight by Charles Lindbergh, in the Ryan monoplane *Spirit of St Louis*, on 20–21 May 1927. As an epic of courage and navigational skill it has never been surpassed.

In the same year, Sir Alan Cobham flew the prototype Short Singapore flying-boat 23,000 miles around Africa. And in 1928 Sir Charles Kingsford Smith, Charles Ulm and their crew crossed the 6,850 miles of the Pacific Ocean, from San Francisco to Brisbane, for the first time in their Fokker monoplane, *Southern Cross*.

Another Fokker had been used by Lt Cdr (later Rear-Admiral) Richard Byrd of the US Navy for the first flight over the North Pole on 9 May 1926. Now, in November 1929, he followed with the first crossing of the South Pole in a Ford Tri-motor, piloted by Bernt Balchen.

One spot on Earth's surface remained to be conquered – Mount Everest, highest of all. This, too, passed under the wings of aeroplanes on 3 April 1933, when the two Westland biplanes of the Houston-Everest Expedition photographed its summit from a height of 34,000 feet.

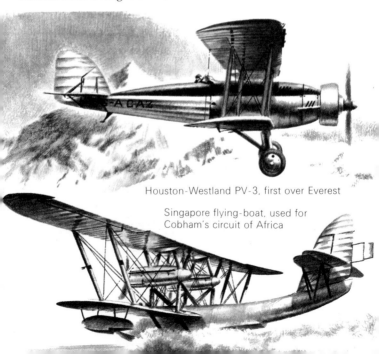

Houston-Westland PV-3, first over Everest

Singapore flying-boat, used for Cobham's circuit of Africa

FLYING BECOMES NATIONALIZED

The first international airlines

It was one thing to fly over the Atlantic in specially-prepared aircraft, and quite another to try to carry passengers over the early air routes. The only aircraft that could be used were converted wartime bombers. There were no proper airports or meteorological services, and once a pilot took off he was on his own until he reached his destination.

Surprisingly, the first post-war services were opened by the recently-defeated Germany. On 5 February 1919, Deutsche Luft-Reederei inaugurated a daily passenger service between Berlin and Weimar, via Leipzig. Its fleet included LVG C.VI bombers, converted to carry two passengers in the open rear cockpit, and mail.

On 8 February, the Farman company staged a cross-Channel flight which did not qualify as a passenger service because those on board were military personnel; but the roomy enclosed cabin of the Goliath aircraft, and food and cham-

pagne served in flight, gave a hint of future airline comfort.

Farman operated the first true international airline service, between Paris and Brussels, a few weeks later, on 22 March. But the most significant date to remember is 25 August 1919, when Aircraft Transport and Travel Ltd opened the world's first daily international service. The aircraft used was a D.H.16, piloted by Cyril Patteson, who took four passengers from Hounslow to Le Bourget. Fare for the 2½-hour flight was £21.

Not all AT & T passengers enjoyed the luxury of an enclosed cabin. Many travelled in open-cockpit D.H.9s, but were supplied with leather coats, helmets, goggles, gloves and even hot water bottles to keep them warm; and things

(*Opposite page, top*) one of AT and T's open-cockpit D.H.9 airliners. (*Left*) Farman Goliath of Air Union, a predecessor of Air France. (*Top*) two of Luft-Reederei's converted bombers: a twin-engined AEG G.V and single-engined AEG J.II, used on the first Berlin-Weimar services. (*Right*) pioneer radio station in a Handley Page transport.

gradually improved. Handley Page airliners even began carrying Marconi radio in 1920, enabling pilots to keep in touch with aerodromes *en route*.

Typical of the aerodromes that were allocated to the pioneer airlines was London's Croydon Airport. At first it was nothing but a large grass field, with a few wooden huts for company offices and wartime hangars for their aircraft. A road called Plough Lane ran across the Airport, cutting off the hangars from the flying field. This led to installation of the first level crossing for aeroplanes – a barrier which was put across Plough Lane to halt road traffic whenever an airliner had to taxi over to the flying field to pick up its passengers.

One of the first meteorological services was that organized by the Dutch airline KLM. Aircraft bound for a Channel crossing were routed via a house belonging to one of the company's employees on the continental coastline, where they circled the garden at low altitude. The pilot then read on a blackboard if the weather over the sea made the trip 'Go' or 'No go' and acted accordingly.

Navigation was uncomplicated. With most journeys made at low altitude, pilots simply carried a map and followed convenient roads and railways. This produced occasional problems – like when a Paris-bound airliner flying along a straight road in France, with the pilot looking down rather than ahead, met a London-bound aircraft flying at the same height over the same road in the opposite direction. The two machines collided head-on.

Generally, though, the standard of safety established by the pioneer airlines was higher than their record of reliability. One memorable flight between London and Paris included twenty-two forced landings *en route*. Fortunately, none of them occurred over the water and a D.H.9 could be put down in any fair-sized reasonably-flat field.

Multi-engined safety came when the struggling private airlines gradually merged into state-subsidized companies like Britain's Imperial Airways, formed on 1 April 1924. Imperial's policy was to fly only multi-engined aircraft, and in 1927 it introduced the first-ever 'named' service – the lunchtime *Silver Wing* to Paris, with meals for passengers.

A nine-passenger D.H.34 of Daimler Hire, 1922

The prototype D.H. Hercules. Later, all Hercules had an enclosed flight deck.

Cabin of a *Silver Wing* Argosy of Imperial Airways

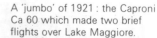
A 'jumbo' of 1921 : the Caproni
Ca 60 which made two brief
flights over Lake Maggiore.

Essays in giantism

Adding a second or third engine did not always guarantee
improved safety. For example, when Alcock and Brown flew
their twin-engined Vimy over the Atlantic in 1919, they
were no more proof against engine failure than were the
unfortunate Hawker and Mackenzie-Grieve with a single
engine of the same type. Indeed, they had to accept twice as
many possibilities of failure and could not have remained
airborne on one engine.

When designers began to suggest the use of multi-engined
airliners in the 'twenties, it was different. Less heavily
loaded than the Vimy, these machines were intended to
maintain height with a full load after an engine shut-down,
and it was this that attracted companies like Imperial
Airways.

The new generation of passenger aircraft ordered in the
mid-twenties were typified by the D.H.66 Hercules (see page
53), which had three 420-hp Bristol Jupiter engines and
could carry seven passengers in addition to two pilots and a
radio operator, plus a large load of mail. It was bought for
use on the hot overseas route to India – the first stage of
which, between Cairo and Baghdad, had been opened up by
the Royal Air Force with an air mail service for troops in
Iraq in 1921. For shorter services, nearer home, Imperial

ordered the three-engined Armstrong Whitworth Argosy, able to seat twenty passengers. By the standards of the time, the Argosy, spanning ninety feet and with a loaded weight of 18,000 pounds, was a 'giant' airliner. A few designers were, however, thinking far beyond such machines, almost to what would be called 'jumbos' today.

No aircraft deserved such a nickname more than the Caproni Ca 60. Its huge hull could seat 100 persons and it had three sets of triplane wings, each spanning 100 feet. It is reported to have flown briefly over Lake Maggiore in early March 1921, but was always viewed more as a research vehicle than a serious airliner.

Very different was Andrei Tupolev's all-metal ANT-20 *Maxim Gorki*, completed in 1934. Powered by eight 850-hp engines, it was unashamedly a propaganda machine, complete with printing press and photo-lab for producing leaflets in flight, loudspeakers for aerial broadcasting, film projection equipment and provision for displaying slogans in electric lights under the wings, which spanned 206 feet. It was lost in 1935, when it collided with a small aeroplane that was performing aerobatics near it, all forty-nine persons on board being killed.

Russia's giant ANT-20 *Maxim Gorki* propaganda aircraft of 1934

Competitors in the Light
Aeroplane trials at Lympne in
1923 included the tiny
Gnosspelius Gull (*above*) and
English Electric Wren (*left*)

The ANEC flew 87½ miles on a gallon of petrol

Minimum aeroplanes

At the opposite extreme in size to the Caproni Ca 60 and
ANT-20 were the aircraft built for the light aeroplane competitions at Lympne aerodrome, Kent, in 1923–26. The first
contest, organized by the *Daily Mail*, was for motor-gliders
that might make flying sufficiently inexpensive for the
general public. Joint winners of one prize, for the longest
distance flown on a single gallon of petrol, were the English
Electric Wren and ANEC I, both of which covered 87½ miles.

Great ingenuity went into the design of these lightplanes.
The Gnosspelius Gull, for example, gave its pilot an excellent
forward view by having its 698 cc engine mounted behind

the cockpit and driving two pusher propellers through chains. Even better was de Havilland's Humming Bird, which the RAF ordered in small numbers to provide inexpensive flying training for its pilots.

None of these aircraft was, however, a really practical flying machine, suitable for hard usage and for flight in anything but good weather. Believing that he knew better the sort of aeroplane that would-be flyers wanted, Captain (later Sir) Geoffrey de Havilland produced the prototype two-seat Moth, which flew for the first time in February 1925.

No aeroplane has exerted a greater influence on flying progress. The production version sold for only £595, could be towed by a car and kept in a garage. It was easy to fly, sturdy and so economical to run that it started a boom in private and club flying that spread across the world.

Many great long-distance flights were made in Moths, none more celebrated than the solo flight from England to Australia made by a young, newly-trained girl pilot named Amy Johnson, in 1930.

Eight de Havilland Humming Birds were bought by the RAF

The D.H.60 Moth made possible the worldwide flying club movement

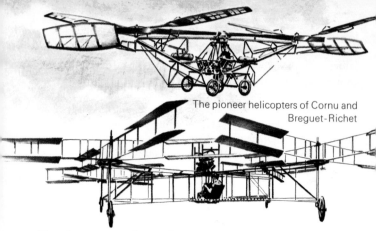

The pioneer helicopters of Cornu and
Breguet-Richet

Getting up and staying up

Throughout the years in which aeroplanes were becoming
more practical and widely accepted, there were men who
believed that progress was following the wrong path. Each
increase in speed and load-carrying ability seemed to be
accompanied, automatically, by higher take-off and landing
speeds and longer take-off and landing runs. Most accidents
occurred during these periods of flight, and it seemed worthwhile to find ways of reducing take-off and landing speeds
without loss of efficiency in other ways.

One obvious answer was the helicopter, which had been
tested successfully in France as far back as 1907. The Breguet-
Richet Gyroplane No. 1 had lifted itself from the ground on
29 September that year, carrying a pilot; but had been held
steady by four assistants. Shortly afterwards, on 13 November, the first free flight with a pilot had been made by the

Cierva C.8R Autogiro of 1927

tandem-rotor helicopter of Paul Cornu; but he lacked the money to continue his experiments.

Igor Sikorsky had built two helicopters in Russia in 1909–10. His subsequent switch to fixed-wing aircraft represented no loss of interest in rotating wings, but an acknowledgement of the fact that much time-consuming and costly research was needed, as well as better engines, before helicopters could become practicable.

It was Juan de la Cierva, in Spain, who made the first major breakthrough. He called his invention the Autogiro because its rotor was turned automatically by the air flowing past it, like the sails of a windmill, instead of being engine-driven. This meant that his aircraft needed a normal propeller for taxying and forward flight, to set the rotor turning; but take-off and landing speeds and runs were greatly reduced.

Cierva discovered at first that when the rotor began turning it tended to tip the aircraft sideways. The reason was clear. The advancing rotor blade on one side added its own speed to the speed of air flowing over it, and therefore developed much more lift than the retreating blade on the other side. The extra lift tilted the aircraft so much that the rotor smashed into the ground and was wrecked.

The answer was Cierva's 'flapping' rotor, in which each blade was hinged at the root so that when it developed extra lift it simply raised itself instead of the complete aircraft.

Another new idea that was to be put to practical use many years later was in-flight refuelling. This was first demonstrated by the US Army Air Service in 1923, when repeated refuellings by a flying 'tanker' enabled Lieutenants Smith and Richter to remain airborne more than thirty-seven hours.

Flight refuelling the D.H.4 of Smith and Richter

Handley Page H.P.42

Airliners like the Fokker F.VIIb-
3m pioneered monoplane design

Peaked caps for pilots

Throughout the 'twenties and into the 'thirties, mail was
often regarded as a more attractive and profitable cargo than
passengers. The urgency of delivery justified high postal
rates and government subsidies to the airlines that carried it.
Sacks of letters could be packed tightly to fill any vacant
space in an aeroplane; they never complained about bumps
in flight, or delays, and did not expect to be fed or pampered
en route.

Nowhere was this attitude of mind more apparent than in
America. Commercial air transport did not really start there
until 6 April 1926. Even then the first service, operated
between Pasco, Washington, and Elko, Nevada, by Varney
Air Lines, was restricted to mail-carrying.

A step towards the huge passenger traffic of today came
when Boeing Air Transport won a contract to carry mail over
the 1,918-mile route between Chicago and San Francisco. To
do the job efficiently, Boeing's aircraft works produced a
fleet of twenty-five Model 40A mailplanes in five months.
Each of these aircraft had a small enclosed cabin for two
persons, so that additional revenue could be earned by
offering the first regular transcontinental passenger service.

The pilot still sat in the open, but had a heated cockpit,
which made conditions more bearable on the long haul in
winter. Even here, changes were on the way, for most of the

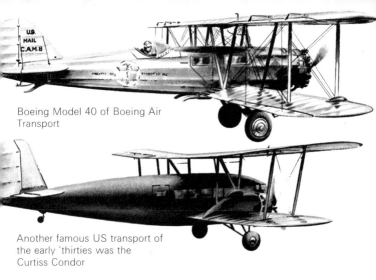

Boeing Model 40 of Boeing Air Transport

Another famous US transport of the early 'thirties was the Curtiss Condor

airliners then on the drawing board were to have enclosed flight decks.

Typical was the Handley Page H.P.42, which began to equip Imperial Airways in 1931. Seeing it at Croydon for the first time, Anthony Fokker, the Dutch designer, referred scathingly to the 'built-in head-winds' caused by its big biplane wings, biplane tail, multitude of struts and massive fixed undercarriage. The H.P.42 was certainly slow, with a speed of under 100 mph; but this gave its passengers time to enjoy a full-course meal in comfort as they cruised majestically between London and Paris. The big wings lifted the aircraft off grass airfields in an incredibly short distance and the 42s were so safe that they were soon carrying more passengers between London and the Continent than all other airliners combined. Yet only eight were built, of which four served on Imperial's Eastern routes between Karachi, Cairo and Kisumu.

Pilots of H.P.42s no longer needed the protection of helmets and goggles. Instead, peaked caps, blue serge uniforms and gold braid became the order of the day. Airliners, too, were changing. Good as it was, the H.P.42 was to be the last of the biplanes and Anthony Fokker's monoplanes were more typical of the next generation of passenger transports.

Heyday of the flying-boat

Before the big monoplane airliners of what we regard as the modern type became standard equipment, there was one more brief, but important, diversion along a different design path.

As more and more of the overseas routes operated by airlines were over water, it seemed logical to utilize a type of aircraft that could use existing harbour facilities and alight on the sea in an emergency. This was illusory, as flying-boats are little safer in the open sea than a landplane airliner that is 'stalled-in' with its wheels up. There were also many instances of flying-boats holing their hulls by hitting objects floating on, or just under, the surface of their landing area; and when the water was absolutely calm it was so difficult to judge height above it that pilots sometimes 'flew in' heavily and sank.

Despite these hazards, for a decade in the 'thirties flying-boats reigned supreme, offering such graceful, luxurious travel that they are still remembered with biased affection.

First of them was Germany's big Dornier Do X, a twelve-engined giant that once lifted a record load of 10 crew, 150 passengers and 9 stowaways but otherwise achieved little but a near-disastrous transatlantic journey. At the time, Imperial Airways used a few flying-boats on the trans-Mediterranean stages of its Empire routes but hardly anywhere else. It was, therefore, quite a surprise when the airline ordered a fleet of 28 Short Empire 'boats.

The gamble was justified. The prototype, *Canopus*, entered

Dornier Do X of 1929,
largest aeroplane of its time

service in October 1936, less than four months after its first flight; and the whole fleet, increased by later orders, not only made the airline the envy of the world but made possible the Empire Air Mail Scheme, by which all first-class mail was carried by air throughout the British Empire without surcharge.

In America, Pan American Airways also built up its reputation with water-borne aircraft, starting with small Sikorsky amphibians and progressing to the big four-engined Sikorsky S-40/S-42, Martin 130 and Boeing 314 *Clippers* with which it pioneered services across the Pacific and Atlantic.

(*Top*) Sikorsky S-40, prototype for Pan Am's famous S-42 *Clippers*. (*Centre*) Short 'C' Class Empire flying-boat. (*Bottom*) Boeing Model 314A taking off.

RETURN OF THE MONOPLANE

Birth of the modern airliner

Paradoxically, the modern landplane airliner was born in the decade when the flying-boat dominated the world's long-distance air routes. Since the early 'twenties, companies like Fokker in the Netherlands and Junkers in Germany had preached the advantages of clean, cantilever monoplane design in a biplane age. Junkers had also advocated all-metal construction, and the two ideas were combined in their sturdy three-engined Ju 52/3m, which served as a bomber in the newly-created *Luftwaffe* and as a 15/17-passenger transport with Deutsche Lufthansa and other airlines.

It was, however, in the United States that the full concept of the modern airliner took shape. Boeing Air Transport and other operators had merged to form United Air Lines and were looking for a replacement for their three-engined Fords, Fokkers and Boeing Model 80s. The aircraft which Boeing built to do the job, the Model 247, was slightly smaller and lighter, 50 to 70 mph faster and could climb on one engine

Douglas DC-3, known in Britain as the Dakota

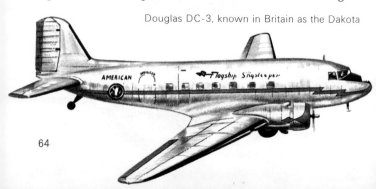

(*Left*) Boeing 247, the first of the 'modern' airliners

(*Below*) the Junkers Ju 52/3m served in both civil and military forms

with a full load of ten passengers. It was all-metal, with a retractable undercarriage, control-surface trim-tabs and an automatic pilot to make the pilot's job easier, and de-icing equipment. And it reduced the transcontinental flight time to under twenty hours.

The Boeing 247 would be better remembered had it not been followed quickly by an even more remarkable family of aircraft from Douglas. First of them was the DC-1, which set a new transcontinental record by flying from Los Angeles to New York in 11 hours 5 minutes. It was the prototype for the DC-2, which was bought by many airlines; but the *real* pace-setter was the DC-3, most famous piston-engined airliner of all time.

Beautiful competitors were produced elsewhere, like de Havilland's 22-passenger all-wood Albatross. The DC-3 out-shone them all, and still served throughout the world in greater numbers than any other airliner thirty years after it first entered service in 1936.

The de Havilland Albatross was one of the most elegant airliners ever built and one of the last of wooden construction.

Flying freight

Before Cyril Patteson flew Aircraft Tranport and Travel's first scheduled service to Paris, on 25 August 1919, there had been an early-morning charter operation over the same route. Piloted by E. H. 'Bill' Lawford, the company's D.H.4A carried the famous payload of one journalist and a consignment of grouse, newspapers, leather and Devonshire cream.

Newspapers continued to form a popular air cargo throughout the early years of airline operations, because nothing is more dead than yesterday's headlines. Another cargo which has always remained profitable and popular is baby chicks. Needing no food of any kind during their first seventy-two hours of life, they require little special care and are sometimes carried thousands of miles, 200,000 at a time, in specially-designed lightweight cardboard 'miniature apartment houses'.

On the whole, though, air freighting built up even more slowly than passenger carrying. This was understandable, as rates were high and cabins and doors were too small.

Occasionally, the problems of small size were overcome by ingenuity, as when somebody in Canada wanted to send a horse by air into northern Quebec. Canadian Airways offered to handle the job with their big Ju 52 seaplane, which had special loading doors for bulky packages. Unfortunately, the horse was taller than the door or cabin. So it was given a sleeping draught, tied to a pallet, pushed aboard, and flown to its destination before it had time to wake up.

Air freighting in America was an offshoot of the well-established air mail services and was taken very seriously. Unlike Croydon, where the manager of one airline used to guard bullion and diamonds by sleeping with them under his bed, even the humblest air freight received the full security treatment. Thus, the first air freight shipped from Chicago by National Air Transport, on 1 September 1927, was loaded under the watchful eye of a guard with loaded rifle, while a representative of American Railway Express ensured that the pilot signed all the necessary shipping forms. The cargo was a 10-gallon stetson hat for comedian Will Rogers.

(*Opposite page*) small cabins and loading doors restricted early cargoes to items like air mail, express packages, crates of chickens — and a drugged horse

Albatross of Imperial Airways by Croydon's famous tower

When flying was an adventure

For people who flew on the Empire routes of Imperial Airways in the mid 'thirties, the pictures on these two pages will bring back countless memories. Air travel then was very different from the modern, highly-organized kind. Passengers were not crammed by the hundred into a pressurized metal tube and whisked from A to B in a few hours. On the contrary, an intercontinental journey was an expedition to be planned for weeks or months ahead, requiring the right dress for night stops at exciting places *en route*, and sometimes bringing passengers in close contact with people and scenes that now seem to belong to a bygone age of the French Foreign Legion and Tarzan movies.

Even the famous old control tower at Croydon Airport, topped by its lattice radio mast, provided a romantic link with the earliest days of flying that is quite missing in the modern red brick and glass-plated edifices.

Overseas, the setting for a night stop in the Middle East was sometimes not an aerodrome at all, but a fort in the middle of a desert peopled by tribes which were still faintly hostile. The aircraft landed on a stretch of hard sand and was taxied into a stockade, while passengers trooped inside the fort, stealing a slightly anxious glance at the armed sentries but recognizing that it would provide an excuse for 'one-upmanship' for months to come, over tea with friends.

The precautions were no mere attempt to add glamour to a long flight. Hi-jacking in the air may be new, but it was vital then to protect aircraft, supplies of fuel and personal belongings from predators, both animal and human.

Equally adventurous were journeys by flying-boat, which seemed to combine the speed of air travel with the luxury of cruising. After the adventure of embarking by boat in a sunny port came the thrill of watching great walls of white spray curving away on each side as the aircraft gathered speed, followed by the seemingly-slow, gentle turns and climbaway across the sea.

Altogether, it was very different from the economy class travel of today, but far more costly.

Argosy at a desert landing site

Calcuttas and Kents in the harbour at Alexandria

From wood to metal

When passengers first began travelling by air from one country to another they flew in aeroplanes built of the same sort of materials as the Wright 'Flyer' of 1903. The basic design had changed. In particular the fuselage was covered-in; but the structure was still wooden and the covering was still fabric. Even when a passenger compartment was enclosed with a lid containing sliding windows, as on Aircraft Transport and Travel's D.H.16s, the aircraft was noisy, and smelly, and draughty.

Some companies, like Junkers, began making aeroplanes of metal during the First World War. In general, however, wood survived as the standard material for all types of aircraft until the latter half of the 'twenties. Air forces were first to insist on the use of metal basic structures for wings and fuselages, even where fabric covering was retained, to ensure the best combination of light weight and strength for the higher-performance combat 'planes that were beginning to replace those of 1914–18 vintage.

Airliners followed closely after the warplanes, and from the early 'thirties most of those that went into large-scale production were of all-metal construction. In fact, airliners such as the Boeing 247, DC-3 and Empire flying-boat were in advance of many military types in having metal skin as well as metal basic structure. By comparison, even early models of the RAF's Hurricane fighter, one of the victors of the Battle of Britain in 1940, had fabric-covered metal wings, and all Hurricanes had fabric-covered rear fuselages.

How good were the all-metal structures built for the airliners of the 'thirties? Perhaps the best answer is that one DC-3 was still giving good service in the 'sixties with 81,535 flying hours in its log book. This meant that it had flown the equivalent of 480 journeys around the Earth at the equator or 25 trips to the Moon and back. It had spent the equivalent of well over nine years actually in the air, worn out sixty-eight pairs of engines and burned eight million gallons of petrol – enough to keep the average motorist going for about 12,000 years.

Despite everything, more than ninety per cent of its airframe was still original. That's how good a metal airliner is!

Designers and engineers had to learn the new skills required for working in metal instead of wood.

The cockpit of the Vickers Vimy in which Alcock and Brown made the first non-stop transatlantic flight

No more 'seat-of-the-pants' flying

At the time of the first airlines and regular air mail services, pilots were said to fly 'by the seat of their pants'. This meant that they relied on personal feelings, sight and reactions rather than instruments – which was natural enough as they had few instruments. If an aircraft began to bank over more and more steeply in thick cloud, they relied on their sense of balance to tell them. If it began to dive, the 'seat of their pants' tended to lift off their cockpit seat to warn them.

Unfortunately, such reliance on bodily senses sometimes proved misleading. Pilots flying too long in cloud became disorientated, felt suddenly that they were banking, over-corrected and went into a fatal spin.

The cockpit illustrated at the top of this page is that of the Vickers Vimy in which Alcock and Brown made the first non-stop flight over the North Atlantic. It is typical of its time, with a single control wheel, like a motor car steering wheel, and a minimum of instruments. Even the brass electrical switches look as if they would be more at home on the wall of a bedroom than in an aeroplane. Yet, with no more than this sparse panel to help them, and a radio that did not work, the two pioneers found their way over the Atlantic.

Such exploits called for superb navigational skill as well as courage, using dead reckoning, astro-navigation and any other favourite techniques the men had evolved by experience. There was, for example, the rather special navigation aid used exclusively by Dean Hill, one of the pilots on America's first mail run. His sector was the so-called 'hell stretch' from New York to Cleveland, over the Allegheny Mountains. If, as happened all too frequently, they were shrouded in fog or cloud, he made a habit of lighting a large cigar immediately he had taken off from the airfield at Bellefonte. He then cruised along happily in the murk, seeing nothing, until the cigar had burned down to a length of about two inches, by which time he estimated that he should be over his destination. Usually he was, and Hill claimed that his cigar was the first really reliable blind-flying instrument.

By the mid-thirties, things were becoming much more professional on the flight decks of commercial airliners. There were proper blind-flying instrument panels, complete with turn-and-bank indicator and artificial horizon, and radio-compasses to tell the pilot where he was, by giving him 'fixes' on two or more ground stations. A receiver aerial was rotated until it gave the precise bearing of each of the transmitting stations in turn. Where the two bearings crossed on the map was the position of the aircraft. It was a great improvement on watching for railway lines and straight roads, especially at night or in bad weather.

The flight deck of a Junkers Ju 52/3m airliner of the late 'thirties

Supermarine S.4

Supermarine S.6 of 1929

Racers point the way to progress

As we have already seen, the gradual switch from biplane to monoplane design gained momentum in the early 'thirties, with the introduction of aircraft like the Boeing 247 airliner. The process might have taken a lot longer but for the lessons learned with racing 'planes.

Most coveted prize of all in the years after the First World War was the Schneider Trophy for seaplanes. Britain won the 1922 Schneider contest with a Supermarine Sea Lion flying-boat; but the aircraft's designer, Reginald Mitchell, knew that it would not be good enough to repeat its success in 1923. He was right. America entered some Curtiss seaplanes, fitted with closely-cowled Curtiss liquid-cooled engines that represented a completely new concept of streamlining for speed.

Mitchell felt certain that, given a free hand, he could do even better. Supermarines agreed to put up the money, and the result was the incredible S.4 seaplane. Nothing approaching its slim streamlined form had ever been seen before. It had cantilever (unbraced) wings of special high-speed aerofoil section evolved at the Royal Aircraft Establishment, and was powered by a 700 hp Napier Lion engine.

The D.H.88 Comet racer won the
1934 England-Australia air race

In September 1925, the S.4 raised the world speed record
for seaplanes to 226.6 mph. In its final trial, one month later,
on the day before the Schneider contest, it developed wing
flutter and crashed. Mitchell was disappointed but made
certain of victory later with the even more powerful S.5.
Consecutive wins by the further refined S.6 in 1929 and
S.6B in 1931 won the Schneider Trophy outright for Britain,
after which the S.6B put the world speed record above
400 mph for the first time.

Was it worth the trouble and expense? Suffice it to say
that lessons learned with these seaplanes enabled Mitchell to
design a little fighter called the Spitfire a few years later,
while the Rolls-Royce aero-engines which powered the
majority of the best RAF aeroplanes of the Second World
War were direct developments of the S.6B's remarkable
2,500 hp racing engine.

Another aircraft which exerted great influence on design
in the 'thirties was the de Havilland Comet. Built specially
for the great MacRobertson Air Race from Mildenhall,
England, to Melbourne, Australia, in 1934, it not only finished
first but brought Australia within three days' flying time of
Britain for the first time. Its sleek form and retractable
undercarriage were reflected within a few years in de
Havilland's Albatross airliner (see page 65) – proving once
again how much could be learned from building racers.

Minimum aeroplanes – Maximum engines

While British designers were producing machines like the Supermarine S.6B and de Havilland Comet, their US counterparts were working on racers of a very different kind. They stopped relying on military biplanes to provide the thrills at their annual National Air Races and began designing and building a series of specialized high-performance monoplanes that seemed to become more bizarre each year.

Centre of attraction at Cleveland in 1929 was the Travel Air Mystery, designed by two young engineers employed by Walter Beech, who was himself to become eventually one of the 'big names' in the lightplane business. The 'Mystery' tag referred to the fact that when the aircraft arrived at Cleve-

The Chevrolet-engined Travel Air Mystery

(*Below*) the 110-hp Cirrus-engined Command-Aire and (*bottom*) Ben Howard's Pete, first to exceed 150 mph on 100 hp

land Airport it was immediately covered by canvas wraps to shield it from prying eyes.

Piloted by Douglas Davis, this machine, with a 400-hp Wright Whirlwind engine, won the important Free-for-All speed contest, proving itself 70 mph faster than the US Army's then-standard single-seat fighters. It clocked 208.7 mph on its fastest lap – the highest speed recorded by a civil aircraft in America up to that time – and was the first machine ever to outfly both the Army and the Navy.

A Chevrolet-engined version of the Mystery won its first race, despite a sick power plant. In the following year, another machine with an air-cooled in-line engine – Ben Howard's little Pete – became the first aircraft of 100 hp or less to exceed 150 mph and earned enough prize money to pay for itself in one season.

Gradually, however, the big air-cooled radial engine became dominant. Airframes got smaller and smaller as engines grew bigger, the limit being reached, or even sur-passed, with the Gee Bee Super Sportsters. Most famous of these was the 800-hp Wasp-engined R-1, in which Jimmy Doolittle won the 1932 Thompson Trophy Race at 252.7 mph. Spanning only twenty-five feet, both Gee Bees of this type later crashed, and interest switched eventually to good design, to achieve maximum speed for minimum power, rather than the dangerous maximum power/minimum aero-plane formula of the early 'thirties.

The 800-hp Gee Bee Super Sportster

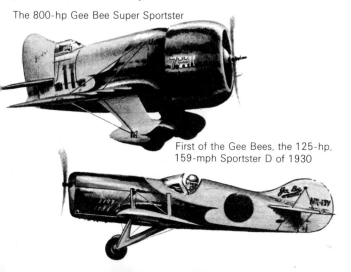

First of the Gee Bees, the 125-hp, 159-mph Sportster D of 1930

New emphasis on research

While the designers of racing 'planes and combat aircraft concentrated on ever-higher speeds, other people devoted their efforts to making flying safer, more versatile or more economical for the airlines.

Accidents from the age-old bogy of stalling – losing flying speed to such an extent that an aircraft dived into the ground – were reduced greatly in the 'twenties by Handley Page's invention of the slotted wing. By mounting a small retractable or fixed slat on the leading-edge, they found that smooth airflow could be maintained over the wing at much slower speeds and far higher angles of attack (wing tilt) than before, and flying became infinitely safer.

Germany's stalky Fieseler Storch was one of the first to demonstrate that the extensive use of slats and large wing flaps would enable an aeroplane to take off and land slowly almost anywhere, setting the fashion for the STOL (short take-off and landing) aircraft of today.

The Russian designer Makhonine developed a different idea in France. His 480-hp monoplane had telescoping cantilever wings, giving high lift from maximum area at take-off and landing, and minimum drag for high speed in flight with the outer wings retracted. It remained only a prototype, but a different kind of 'variable geometry' is used in the 'swing-wing' military aeroplanes of today.

Other people were content simply to utilize large fixed wings for high lift. With 568 square feet of wing area and a highly supercharged Pegasus engine, the Bristol 138A set up a new aeroplane height record of 53,937 feet in June 1937, its pilot garbed in an early version of a modern space-suit. It was a pure research aircraft, but other large-winged types were intended for commercial use. Even the fuselage of the Burnelli transport was given an aerofoil (wing) shape to enable it to contribute lift and so make possible the carriage of heavy loads. But such tapering sections are not ideal for housing people or freight containers.

(*Opposite page, top to bottom*) the Bristol Type 138A high-altitude monoplane ; Fieseler's much-slotted and flapped Storch ; the Burnelli O A 1 flying-wing transport built by Cunliffe Owen in 1938 ; and the telescopic-wing design of Makhonine.

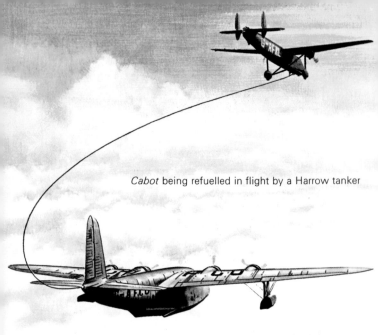

Cabot being refuelled in flight by a Harrow tanker

Flying tankers, catapults and piggy-back 'planes

By the mid-thirties, there had been quite a number of successful flights over the Atlantic, even by light aircraft piloted by superb navigators like Jim Mollison. Zeppelin airships were operating a regular service between the old and new worlds, and airlines like Imperial Airways and Pan American were re-equipping with fine modern flying-boats that seemed good enough to tackle any route, anywhere.

Range was the only problem. The standard Empire 'boat could fly only 760 miles with a full load. *Caledonia* and *Cambria* were delivered unfurnished to save weight, and had six extra fuel tanks installed in the wings and across the top of the hull to increase their maximum range to 3,300 miles; but even then performance was marginal for the crossing to New York with any sort of passenger payload.

So, although *Caledonia* made the first transatlantic flight by a commercial transport aircraft on 5 July 1937, simultaneously with the Sikorsky S-42 *Clipper III* of PanAm, neither carried a payload and it was decided to experiment with a new aid to long-distance flying for future passenger services.

Since the first, rather crude American experiments with in-flight refuelling in 1923 (see page 59), the technique had been developed to a fine art by Flight Refuelling Ltd, a company formed in Britain by one of the nation's most famous pioneer long-distance flyers, Sir Alan Cobham. Consequently, Imperial Airways decided to have four of their Empire flying-boats delivered with equipment for flight refuelling installed and tanks large enough to hold 2,500 gallons of petrol. Taking advantage of the fact that an aeroplane can fly with a bigger load than it can lift off the ground by its own power, these aircraft were designed to take off at a loaded weight of 46,000 pounds but to fly at up to 53,000 pounds after refuelling in the air.

Despite the in-bred reluctance of many airline pilots to fly close to another aeroplane, the first trials went well, and on 5 August 1939 Captain Kelly Rogers opened a weekly trans-atlantic mail service in *Caribou* after refuelling in flight. Only eight crossings were made each way before the Second World War brought the operation to a halt and prevented the addition of passengers to the mail carried on these flights.

The Ha 139 seaplane *Nordwind* and its depot ship

As an alternative to flight refuelling, Imperial Airways experimented with an even more revolutionary idea in 1938 – again based on the concept that an aeroplane can fly at a greater weight than it can lift off the ground. Devised by Major Bob Mayo, the airline's General Manager (Technical), it entailed carrying a small four-engined high-performance mailplane named *Mercury* into the air on the back of the modified Empire 'boat *Maia*. *Mercury* could then be released in the air, laden with far more fuel than it could have lifted unaided.

The Mayo Composite worked. Piloted by Don Bennett, *Mercury* set up a long-distance record which has never been beaten, by flying non-stop 5,998 miles from Dundee, Scotland, to the Orange River, South Africa. It also made the first commercial flight across the North Atlantic, with 600 pounds of mail, on 21 July 1938. But it had no future as a passenger-carrying aircraft.

Meanwhile, the Germans, too, had been flying mail across the Atlantic to New York by aeroplane for some years, by catapulting seaplanes from depot ships in the Azores. They had begun experimenting with such a service as early as 1929, when a Heinkel He 12 seaplane was catapulted from the new luxury liner *Bremen* during the latter's maiden voyage to New York. It flew only the last 248 miles; but on the return trip was catapulted off Cherbourg and arrived in the home

The Mayo Composite taking off

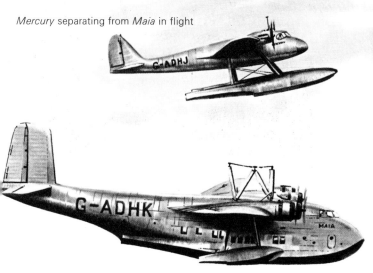

Mercury separating from *Maia* in flight

port of Bremerhaven twenty-four hours ahead of the liner.

By the following year, the longer-range He 58 seaplane was able to save thirty-six hours in the delivery of mail carried by the liner *Europa*; and the service reached its zenith with the construction of special depot ships for carrying and catapulting big Blohm & Voss Ha 139 seaplanes. Between 1937 and 1939, these aircraft made many mail flights across both the North and South Atlantic. But all such ideas were make-shift. The real answer lay in the bigger and entirely conventional Boeing 314 flying-boats with which PanAm opened the first scheduled transatlantic passenger service by heavier-than-air craft on 28 June 1939.

Flying in comfort

Nobody can pretend that aeroplanes have ever provided the same standards of spacious comfort as the rigid airships of the 'thirties. These huge craft could offer their passengers individual cabins with bed, wardrobe and wash-basin, a proper dining room, lounge, dance floor and promenade decks giving a wonderful view of the ground or sea through huge windows.

As aeroplane flights became longer, it was essential to

Sleeping berths in a Latécoère 631 flying-boat, delivered to Air France in 1947

improve the standards of comfort which they, too, offered to passengers. This was less easy, for every 200 pounds of beds, tables, cutlery, curtains, food and water for washing, and every steward or stewardess carried to look after their needs, meant that one fewer fare-paying passenger could be accommodated. This may not matter much today, when an airliner with half its seats filled can still make money by hauling huge loads of freight in its cargo-holds; in the 'thirties profits were often so precarious that every scrap of payload mattered.

It was not, of course, a time of economy class cheap fares or inclusive-tour holiday traffic. Most air travellers were, and were treated as, first class passengers. If they wanted bunks in which to sleep these could be provided, with a higher fare

Lunchtime on board a Junkers G 31 airliner of the late 'twenties

Boeing Model 307 Stratoliner, a pressurized transport development of the B-17C Flying Fortress bomber

offsetting the reduction in passenger payload. Even then, fares could be kept at a reasonable level, because people who slept in the air did not need to be looked after on the ground at night stops. Also, aeroplanes could be kept in the air for more hours of each day when improved instruments, radio and navigation aids permitted regular flying at night and in poor weather. This meant that they spent more of their life earning money and so became more profitable to operate.

Discriminating passengers began to choose the airlines which offered them the highest standards of comfort, safety and service. Later, speed was to become almost the overriding factor in choice of operator, but at this period airliners were seldom really fast and a few extra miles per hour made little difference, especially on short routes.

Then, in 1940, a development of tremendous importance slipped onto the scene almost unnoticed by most air travellers. The Boeing Stratoliner operated by US domestic airlines was the first airliner with a pressurized cabin for flying at high altitudes. In a piston-engine age, there was no point in high pressurization, for the Stratoliner's best cruising height was still only 15,000 feet; but even this was above many of the 'bumps' that created airsickness, and passengers liked it.

Airliners for shorter routes

Not everyone wanted to fly the Atlantic, or cross the world to Australia. In many parts of the world, especially those less well served with railways and roads, people looked to air

Favourite short-range airliner for many years was the D.H. Rapide

travel to carry them to places that were quite near but difficult to reach. So, side-by-side with the big trunk-route airliners, the manufacturers produced small types for operation on local services, where payloads were smaller.

From the start, such aeroplanes came in a great variety of shapes and styles. For the North American market, where speed was rated highly from the start, Lockheed produced a remarkable series of very clean wooden monoplane airliners. Typical was the six-passenger Orion of 1931, with a retractable undercarriage, which could cruise at 200 mph on 550-hp.

Its successor by the mid-thirties was the ten-passenger all-metal Lockheed Electra, with two 450-hp Wasp Juniors and a cruising speed of 185 mph. But on short routes, where

The Orion was a six-passenger all-wood transport of 1931

Successor to the Orion was the all-metal Lockheed Electra.

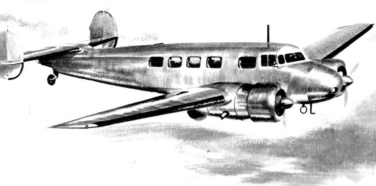

small grass airfields were the order of the day, the lightly-loaded wood and fabric D.H.89 Dragon Rapide biplane, with 200-hp engines, six or eight passengers, and a 132-mph cruising speed was both adequate and extremely economical.

Equally valuable were the aircraft built for 'bush' flying in places like Northern Canada. Operating on skis in winter, wheels or floats in summer, these machines introduced the speed and convenience of air travel to places where life had previously been precarious and sickness often needlessly fatal because a doctor was too far away. Like the old Fairchild monoplane illustrated below, they were seldom elegant in appearance, but they brought flying right to people's back doors at a time when the bigger airliners were beginning to depend increasingly on large and specially-equipped airports.

Fairchild F.C.2 'bush flyer' in the far North of Canada

Aerial workhorses

The executive or business transport is a comparatively recent innovation; but it had a prototype in the Supermarine Air Yacht built for the Hon. A. E. Guinness in 1929. Powered by three 485-hp Armstrong Siddeley Jaguar engines, this big machine carried a crew of three and only six passengers, and hinted at the standards of comfort and speed that the executive aeroplane would one day offer.

Two years before it was built, an Australian padre named John Flynn initiated the now famous Royal Flying Doctor Service by hiring a D.H.50 aircraft and crew from the Qantas airline and appointing a doctor to attend urgent medical and accident cases and to supervise the movement of patients to hospital when necessary. Allied to small radio transmitters that could be operated by a housewife, the service quickly spread a mantle of safety over the 'outback'

The Hon. A. E. Guinness's Supermarine Air Yacht taxying past his seagoing yacht, the *Fantôme*

G-AASE

Stearman Model 75 crop-sprayer in action

that made it seem tolerable and far less remote.

Such services are taken for granted today, as are the survey aircraft which assist town planners and roadmakers, and search for minerals in places where there are no roads; and agricultural aircraft which spread or spray chemicals to kill insect pests and make barren soil fertile. Yet no other vehicle can perform these tasks so well, or even at all. Nothing is more vital than finding new sources of raw materials and increasing the production of food in a world where the population is increasing with such frightening rapidity, and the men who fly these 'aerial work' machines are an *elite* band of specialists. An air ambulance would be little use if it operated only between nine and five o'clock in good weather, and it takes a special kind of courage and skill to fly a crop-duster a few feet above the ground, in hilly or tree-speckled country, with quick, steep turns between each run.

Special survey equipment on this Piper Aztec enables it to detect mineral deposits in the earth
(*Left*) patient inside the cabin of the de Havilland Drover air ambulance (*below*) of the Royal Flying Doctor Service of Australia

Spectacle and excitement were the keynotes of pre-war 'air circuses' and local air displays, and aircraft like the tailless Westland Pterodactyl IV and Cierva C.30 Autogiro kept the public informed about new research trends.

Getting people air-minded

Today air travel is a part of everyday life, and we forget how difficult it was to make the public air-minded in the years between the two World Wars. Military displays, like the annual RAF Pageants in Britain, gave the public an interest in flying but did little to make the spectators want to fly themselves. The barn-stormers achieved more, by touring with their war-weary aeroplanes, offering joy-rides wherever they could find a field and some willing passengers. Not until the first big 'air circus' was organized by Sir Alan Cobham did anyone really get down to the task of

enticing ordinary people into the air in huge numbers.

The formula was simple. Posters announced, well before the chosen date, that the 'circus' was coming to town. On the big day, a large field was roped off, a public address system was set up in one corner, and then the aircraft arrived, drawing vast crowds to the temporary airfield.

They were usually a mixed bag with, perhaps, an old ex-Imperial Handley Page W.10 to carry joy-riders at five shillings a time; veteran Avro 504s and a Spartan Three-seater for more personal 'flips' with a loop thrown in for those who could afford it and were brave enough; a Lincock fighter prototype for high-speed (180 mph in a dive!) aerobatics; and any oddities that could be collected together, such as the tailless Pterodactyl or a Cierva Autogiro.

Attractions included parachute jumps, bombing a car with bags of flour, and shooting from the air at a row of balloons (which were popped by a gentleman with a knitting needle, hiding behind a sacking screen). But the real business of the day was to get as many people as possible off the ground. On a good day, the old Avros were kept so busy that they were still taking off and landing at dusk, when, in the dim light, it seemed that one could see the tired pistons going up and down inside red-hot cylinders.

Geoffrey Tyson inverted in a Tiger Moth

Personal 'planes

Star of many of Alan Cobham's 'circuses' was a Tiger Moth biplane flown by a young man named Geoffrey Tyson. He used it to pick up a handkerchief from the ground with a hook attached to one wingtip, and flew it upside down under a line of bunting between two poles. He specialized in inverted flight, and on the twenty-fifth anniversary of Blériot's first Channel crossing celebrated the event by flying from England to France upside down.

Moth biplanes also continued to form the basic equipment of flying clubs, and an increasing number of people began to buy their own lightplanes, especially in America. When they did so, they soon grew tired of sitting in open cockpits, exposed to the slipstream in all weathers, and looked for something more comfortable.

It was to meet this need that de Havilland produced the Puss Moth in 1930. Unlike the original Moth this was a high-wing monoplane with a steel-tube fuselage. The well-proven 120-hp Gipsy engine, designed specially for the Moth in 1927, was retained, but the pilot and his one or two passengers now enjoyed the luxury of an enclosed cabin. It was all rather civilized; but the Puss Moth was no family car of the air, suitable only for pottering from one local airfield to another.

On 18 August 1932, Jim Mollison took off in his Puss Moth, *The Heart's Content*, from Portmarnock Strand, Dublin, and flew it across the Atlantic to Pennfield Ridge, New Brunswick, in 31 hours 20 minutes, at a cost of £11:1:3d for petrol and oil. Both he and his wife, Amy Johnson, flew aircraft of this type from England to the Cape in record times. Later, he took *The Heart's Content* all the way from England to Natal in Brazil, becoming the first person to cross the South Atlantic solo from East to West. The little aircraft was also the first to conquer both the North and South Atlantic.

Meanwhile, over in America, another little enclosed-cabin aircraft had entered production in 1931. Known at first as the Taylor Cub, after its designer, it was to be taken over later by W. T. Piper and establish the reputation of what became one of the world's greatest lightplane companies.

Popular lightplanes of the 'thirties were the B.A. Swallow of German origin (*top*), Taylor Cub (*centre*) and de Havilland Puss Moth (*bottom*)

Wooden structure of a typical *Pou-du-Ciel*

Main parts included a tilting front wing.

An early Flea taking off. Engines ranged from 22 to 38 hp.

The saga of the Flea

There has never been anything else quite like the Flying Flea or, to give it its correct name, the *Pou-du-Ciel*. The prototype was built in France by Henri Mignet, who wrote a book describing its design, construction, and the fun he had received from flying it. Monsieur Mignet was such an enthusiast, and conveyed so vividly the pleasure of building and flying one's own little aeroplane, that before long people were constructing Fleas all over the world.

Alas, they did not experience the same success as Mignet. Lacking all knowledge of how aeroplanes should be built, they used unsuitable materials, changed the design in important respects, and tried to take off with any old engine on which they could lay hands and which seemed to fit. In vain, the designer warned them to be more careful and to follow his instructions implicitly. Not since Santos-Dumont had offered the public the 'do-it-yourself' Demoiselle prematurely had anyone conceived such a simple aeroplane, and before long a number of people for whom the simplicity was deceptive were lying dead in the remains of their Fleas.

In Britain, a great 'Flying Flea Display' was organized to demonstrate the potential and safety of the design. Completed and half-completed aircraft converged on the airfield and soon the air was filled with the raucous sound of tiny engines struggling to become airborne. Fleas hopped across the grass in all directions; one succeeded in becoming airborne, only to end up in a tree – it was, ironically, a clipped-wing trainer model that was not intended to fly.

Meanwhile, over in France, the authorities had put a *Pou-du-Ciel* in the wind-tunnel and had discovered that if it entered a dive at an angle of more than fifteen degrees to the horizontal, the nose could not be raised to prevent a crash. Mignet continued to protest the innocence of the Flea and to fly his own version safely. Eventually, after the Second World War, he produced refined models, still with the original tandem wings and fore-and-aft control by tilting the front wing. They flew safely, and today there are many Fleas and Flea-types all over the world, giving their owners the kind of happy, economical flying of which Mignet knew his little aeroplane was capable.

Vampyr glider of 1921 in Germany

Silent wings

Attempts by Santos-Dumont, Henri Mignet, the organizers of the 1923–1926 Lympne Light Aeroplane Competitions, and others to find a formula for low-cost 'flying for all' were unsuccessful. Was there any other way of getting people into the air at the sort of price they could afford? One answer was born out of the Versailles Treaty concluded between the victors and vanquished after the First World War.

Young Germans, forbidden to practise any form of powered flying except on airline services, turned to gliding as an outlet for their thwarted ambitions. Although progress had been made since the days of Lilienthal's 'hang-gliders' the best aircraft of the time were quite primitive. The best glide that any of the twenty-four sailplanes gathered together for the first Rhön meeting in 1920 could achieve was a distance of just over one mile in a time of 2 minutes 22 seconds, by a steady descent over a drop of 1,080 feet.

Typical early gliders of 'boxy' design

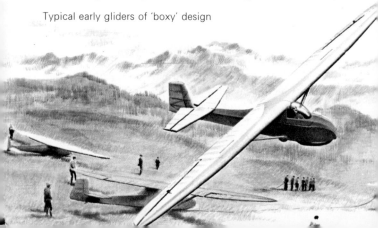

By 1922, gliders like the Greif were beginning to look very like their modern counterparts, although there was still much to learn about wing design and the ways in which flights could be prolonged by the use of natural air currents. Not for some years was it appreciated that soaring was possible by flying in the up-currents along the crest of a slope. Later came a knowledge of 'thermals', rising columns of warm air in which a tightly-turning sailplane could actually gain height and then fly long distances cross-country by hopping from one thermal to the next.

There were considerable differences at first between a 'glider', which simply made a descending flight after launch and was used mainly for training, and a 'sailplane' which was capable of soaring flight. Gradually, however, the trainers improved to such an extent that the two words became almost synonymous.

Launching techniques evolved from the original tow-line, hauled by two columns of energetic helpers, to the modern systems of a motor-driven cable-drum and aircraft towing. No great sums of money were needed; instead many members 'earned their keep' by helping to launch other people until their time came to sit at the controls for the breathtaking climb, gloriously silent descent and, if they were really skilled, the graceful aerobatics and long-distance flights that became the hallmarks of the skilled pilot.

The German Minimoa, one of the most graceful sailplanes of its time

Another pre-war German design, the Olympia

The first practical helicopters

Not until twelve years after Juan de la Cierva perfected the flapping rotor (see pages 58–59) did anyone succeed in building a successful helicopter. Even then both machines which flew, in France and Germany, used a pair of opposite-rotating rotors, to cancel out the rotor torque which tended to spin a single-rotor helicopter on its own axis. This complicated their structure and made them so heavy that neither could carry a worthwhile load.

First to fly, on 26 June 1935, was the Breguet-Dorand Gyroplane Laboratoire, with co-axial two-blade rotors. It looks crude by today's standards, but represented a tremendous step forward. At that time the world helicopter records, established by an Italian d'Ascanio machine, stood at a distance of about two-thirds of a mile, altitude of 59 feet and endurance of 8 minutes 45.2 seconds. On 26 September 1936, the Breguet-Dorand climbed to 518 feet. A few weeks later, it flew for 1 hour 2 minutes 5 seconds and set up a speed record of 27.77 mph over a straight course.

These records soon fell to an even more advanced aircraft

Sikorsky's 1910 helicopter (*left*), Breguet-Dorand Gyroplane Laboratoire (*top right*) and Focke-Achgelis Fw 61

from Germany – the Focke-Achgelis Fw 61. This flew on 26 June 1936, and had two side-by-side opposite-rotating three-blade rotors. It was the first helicopter to make an engine-off autorotative landing, in May 1937, and in the next two years set up a string of records including an altitude of 11,243 feet and distance of 143 miles in a straight line.

The Fw 61 was sufficiently promising for a large transport helicopter on the same lines to be evolved in the Second World War; but by that time there was again something far better in existence.

This time the designer was Igor Sikorsky, who returned to helicopters after nearly thirty successful years as a builder of fixed-wing machines. His aircraft was designated VS-300 and was powered by a 75-hp Lycoming engine. It consisted of an open steel-tube fuselage structure, with a single three-blade main rotor and a vertically-mounted tail anti-torque rotor, and made its first tethered flight on 14 September 1939, piloted by its designer. Free flights began on 13 May 1940, by which time two horizontal tail rotors had been added, on outriggers, to offset temporary control problems.

It continued in use as a test vehicle for two more years, undergoing frequent design changes. By then, Sikorsky had in production the improved R-4, the aircraft upon which the entire world helicopter industry was founded.

The pioneer Sikorsky VS-300 hovering in 1940

TURNING POINT

Radar developed in the Second World War

No British interceptor fighter at the start of the war had a bump on its nose like that of the Beaufighter in the picture opposite, or aerials like those on the German Bf 110. And no bomber had a blister fairing under its fuselage, like that of the Lancaster at the bottom of the page.

In 1939–40, the usefulness of radar was limited to transmitting signals from ground stations to locate and track incoming enemy raiders, so that fighters could be directed by radio to intercept them. This was sufficient to ensure for RAF Fighter Command victory in the daylight Battle of Britain. However, when the *Luftwaffe* sought the cover of darkness for its raids, in the autumn of 1940, it was necessary for the fighters to carry their own radar to track down the enemy in the night sky.

Early British airborne radars (AI) used a number of small separate transmitting and receiving aerials. Later, the complete aerial system could be put into a nose 'thimble' like that on the Beaufighter. The *Luftwaffe* passed through a period of using huge aerial arrays, like those on the Bf 110, before also perfecting a 'thimble' type.

The radar under the Lancaster, known as H2S, produced an electronic map of the ground in front of and below the aircraft, enabling the radar operator to assist in navigation and to find the target with great accuracy at night and in the worst weather conditions.

When peace returned, such devices were at the disposal of the post-war airlines. They opened up the prospect of enormously improved navigational accuracy in all weathers. Nor was this the limit of radar's usefulness. By transmitting signals forward of the aircraft, it was able to detect any high ground which might present a collision risk. It could even detect the centres of areas of storm or dense cloud ahead, enabling the crew to fly around them.

Different types of Second World War radar included the Beaufighter's AI Mk VII (*top*), Bf 110's FuG 212 (*centre*) and Lancaster's under-fuselage H2S

Concrete runways

Equal in significance to the development of radar aids was the wartime construction of runways throughout the world. Up to 1939, even major international airports like London's Croydon were hardly more than large grass fields, with attendant terminal buildings and hangars. Only when heavily-loaded bomber and transport aircraft had to be sent into action day after day did it become essential to build proper hard runways to support their weight in all weathers.

Fighters could often manage with lengths of pierced steel planking, laid down temporarily in combat areas. The bigger aircraft needed concrete by the mile. The techniques and equipment used by engineers to meet the requirements in the shortest possible time were used after 1945 to speed the construction of runways in places by-passed by the war.

The main effect of all this runway construction was to reverse the pre-war trend away from landplanes to flying-boats for long-distance airline services. An operator like British Overseas Airways Corporation (BOAC), which had taken over from Imperial Airways and acquired the latter

Handley Page Halifax III bombers taking off from wartime-built concrete runway

company's flying-boats, now found itself at a disadvantage. With most of its competitors equipped with landplane airliners, bought at low cost from war-surplus stocks, it was often alone in operating flying-boats to many destinations overseas. This meant that it had to cover the whole cost of maintaining marine bases, the boats needed to service its aircraft and carry passengers to and fro, fire tenders, and means of lighting the waterways and keeping them free from floating debris.

The introduction of complex new radio and radar aids only made matters worse. It was unthinkable to install expensive 'talk down' and instrument landing systems at a marine base used only occasionally by one airline; yet safe, on-time services demanded such equipment.

Competing airlines flew their landplanes into airports ashore, provided, paid for and staffed by local authorities, paying only landing fees to cover each flight. This was far less expensive, and BOAC knew that eventually it would have to dispense with its flying-boats, popular though they were, and switch to an all-landplane fleet like everyone else.

A wartime generation

In the autumn of 1945, it was clear that the aircraft industries of the world would need time to switch from the mass production of combat aircraft to the development and construction of civil airliners and lightplanes. With commendable foresight, the British government had set up a committee to report on future commercial requirements while the war was still being fought. Chaired by Lord Brabazon, holder of the first UK private pilot's licence, it recommended a whole series of new machines, one of which, the Type IV, was to take advantage of Britain's unrivalled experience and leadership in the new field of jet propulsion.

This made sense. Under wartime agreements, the USA had supplied all the transport aeroplanes needed by the western allies. This led to the production of more than ten thousand DC-3/C-47s, and the setting up of massive assembly lines of four-engined DC-4/C-54 Skymasters, Constellations and other types. No other nation could compete with such large-scale manufacture of war-developed and financed piston-engined aircraft, and it was clearly better for Britain to make do with ex-military and interim machines for a few years, while taking a great leap forward into the jet age.

This produced some strange scenes. Visitors to Croydon Airport, for example, were treated to a vision of ex-enemy Junkers Ju 52/3m tri-motor transports in service with British European Airways (BEA), newly formed to take over the domestic and European services of BOAC.

As might be expected, ex-military C-47s were snapped up by most of the world's airlines, as ideal aircraft with which to re-establish their shorter services. They were to prove so

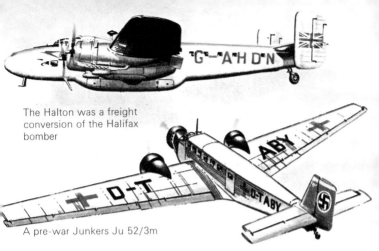

The Halton was a freight conversion of the Halifax bomber

A pre-war Junkers Ju 52/3m

reliable and so economical that they would continue to outnumber any other type for more than two decades.

Then there were the transport adaptations of military designs. Some, like the Handley Page Halton, were simply bombers with a freight pannier built on to their former weapons-bay, as a quick, economical method of utilizing existing ironmongery. But for the fact that the basic airframes were cheap and immensely strong, they were not particularly suitable for commercial use and soon disappeared from the scene. Aircraft like the York, with a completely new and roomy square-section fuselage built on to the basic wings, tail unit, undercarriage and engines of a Lancaster, were sufficiently new to last longer in commercial service; but it was clear that the airlines would need new machines quickly, and in large numbers.

(*Opposite page*) the Avro York was a transport development of the Lancaster bomber
(*Below*) best-known of all, a Douglas C-47 (DC-3) with its freight doors open

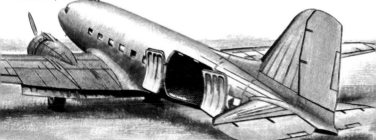

Using the basic wings and other components of the Wellington bomber (*above*), Vickers evolved the Viking (*below*) for the newly-formed British European Airways.

Using bomber 'know-how'

Re-equipment of the airlines tended to follow a three-stage pattern after the Second World War. Stage one had entailed simply painting civil markings on ex-military machines like the Ju 52/3m, C-47 and York and furnishing them as comfortably as possible, as quickly as possible. There was no time for elaborate rebuilding, and it was accepted as normal for a thinly-disguised Liberator bomber to land at a major airport and discharge from its former bomb-bay passengers who had flown hundreds of miles in conditions that are best described as fast rather than luxurious.

Stage three was to bring entirely new and advanced designs. Meanwhile, even the 'stage two' generation of postwar transports was very different from such minimal conversions. In designing their twin-engined Viking airliner, Vickers utilized the fabric-covered outer wing panels, undercarriage and Hercules engines of the Wellington bomber, allied to a tubby, entirely new fuselage seating twenty-one passengers in real comfort. The prototype flew on 22 June 1945, before the war in the Pacific had ended. As a result, it was possible for Vikings to begin work on BEA routes just over one year later, in August 1946. Eventually, this airline had sixty-four of them, all but the first few with metal-covered wings. They remained in service until 1954, flying a total of sixty-five million miles, carrying 2,748,000 passengers and earning £35 million in revenue.

It is not too much to say that the great reputation which BEA acquired early in its life was almost entirely due to the Viking, which carried thirty-six passengers in its 1952 'Admiral' class configuration. It brought London-Paris fares down to £9.15s return, and displayed its great strength when an explosive device planted inside the baggage compartment of G-AIVL almost severed the tail unit. With superb courage and skill, Captain J. Harvey brought the aircraft and its thirty-one occupants safely home to London's Northolt Airport, and the Viking was soon back in service.

Another great post-war transport, based on a bomber design, was the Boeing Stratocruiser. It married the wings, undercarriage and engines of the four-motor B-29 Super-fortress bomber to a new two-deck pressurized fuselage. Most of the lower deck was set aside for baggage and freight; but it also housed a small cocktail bar, approached by spiral staircase from the main cabin. This made the Stratocruiser a firm favourite with passengers on the transatlantic services that were accepted as routine after vast wartime air ferry operations had shown what was now possible and safe.

The luxurious Stratocruiser airliner (*below*) was a two-deck development of the Boeing B-29 atomic bomber (*above*)

Last of the old giants

In only one category of aircraft did the British Brabazon Committee guess wrongly. When considering future long-range operations, it tended to think in terms of the spacious luxury of pre-war first class services, not appreciating that the higher speed and necessarily lower fares of mass air travel would lead to passengers being packed ever more tightly into the airliners of the 'fifties and 'sixties. The miscalculation was understandable, as nobody could have foreseen that shorter journey times made practicable by jet propulsion would make 'human sardine' layouts tolerable.

So, to meet the anticipated needs of British intercontinental air travellers, two enormous anachronisms took shape in a specially-erected 'world's biggest' hangar at Filton, Bristol, and by the slipway of the Saunders-Roe flying-boat assembly workshops at Cowes.

First to fly, on 4 September 1949, was the Bristol Brabazon landplane, weighing nearly 130 tons, designed to carry 100

G-ALUN, the beautiful Princess that nobody wanted

passengers for 5,500 miles at 250 mph, powered by eight 2,500-hp Centaurus piston-engines (but planned to use turbo-props in its production form), and costing the then-unprecedented sum of £3 million. It was mighty, magnificent and as outdated as the dinosaur. What it showed, eventually, was how much better the same job could be done by the far smaller turboprop Britannia (page 126).

The Saunders-Roe Princess flying-boat was even more beautiful and only a little smaller in overall dimensions. Its ten 3,780-hp Proteus turboprops, eight of them in coupled pairs, were intended to give it a maximum range of 5,270 miles at 360 mph, carrying up to 200 passengers on two decks. It was, therefore, a more practical transport than the Brabazon, but unwanted in a landplane age. Cocooned and beached for years, the prototype and its two nearly-completed sisters were finally towed off to the scrapyard, which had long ago broken up the Brabazon.

Britain was not alone in building such monsters of the past. In America, millionaire Howard Hughes dreamed up the Hercules, an all-wooden flying-boat which spanned 320 feet and remains the largest aeroplane ever flown. Piloted by Hughes himself, it made one 1,000-yard hop off Long Beach Harbour in November 1947, at a height of 70 feet, but never flew again. There were to be later giants, but they would have jet engines, more compact all-metal structures and seats for up to 490 passengers.

Largest aeroplane ever flown, the Hughes Hercules of 1947

Douglas DC-4 . . .

. . DC-6

. . . and DC-7

Last of the big piston-engined transports

The days of the big four-engined piston-engined airliner on first class services were numbered as the 'forties gave way to the 'fifties. There was to be just one final generation of superb aircraft from the USA, owing much to wartime transport designs but embodying every new idea to improve comfort and service – except the most important of all, the jet engine.

From the DC-4/C-54 Douglas evolved first the DC-6 family. They were larger and more powerful, with 2,400/2,500-hp Pratt & Whitney R-2800 engines and pressurized cabins. First of them was the basic DC-6, with a length of 100 ft 7 in, which flew for the first time on 15 February 1946. It had, in fact, been developed as the XC-112 to meet military requirements, and 167 cargo-carrying versions, with large side-loading freight doors in a 5-feet-longer cabin, were bought by the US Air Force and Navy. Similar DC-6A Liftmasters entered airline service side-by-side with the DC-6s.

By 1951, Douglas had lengthened the cabin by a further 13 inches to produce the DC-6B, which could carry from 54 to 108 passengers in diminishing standards of comfort. It was a sound, reliable transport, and 286 were sold to the airlines. This encouraged Douglas to 'stretch' the basic design further to produce the DC-7 family, with 3,400-hp Wright Turbo-Compound engines, although the jet-powered Comet was already heralding the new era of jet travel on the European side of the Atlantic.

Last of the breed was the DC-7C, which was so good that even BOAC bought ten to tide it over the gap between the unfortunate Comet 1 jet-liner (page 127) and the Comet 4. The 'Seven Seas' could carry up to 105 passengers 4,100 miles at 274 mph, and ended its life as the DC-7F freighter.

Companions and competitors to these last piston-engined DCs were Lockheed's Constellations, Super Constellations and Starliners. Like the DC-4, they had begun their service in wartime, as military C-69s, and the first 'Connies' bought by airlines post-war were, in fact, made up from components of cancelled military machines.

As in the case of the DC-6/DC-7 series, the basic design

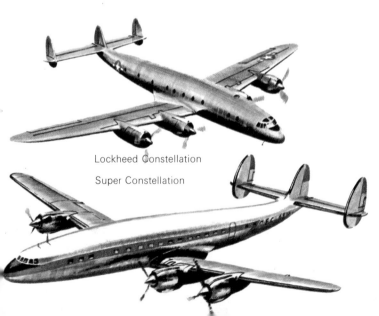

Lockheed Constellation

Super Constellation

underwent a continuous process of 'stretching' and re-stressing to take more powerful engines. Thus, the first commercial Model 049 Constellations had 2,200-hp Wright Cyclone engines, a loaded weight of 86,250 pounds and carried 43 passengers. The final Model L-1049H Super Constellation had 3,400-hp Turbo-Compounds, a weight of 137,500 pounds and seats for up to 109 passengers when it was not used for cargo carrying.

To earn the last possible scrap of revenue from a well-tried and popular design, Lockheed began evolving the L-1649A Starliner in 1955. Trans World Airlines wanted an aircraft for its transatlantic services, so the basic Super Constellation fuselage was fitted with a completely new square-tipped wing housing 8,000 gallons of fuel. The first Starliner flew on 10 October 1956 and only forty-five were built, for Lockheed were already working on their first turbine-powered airliners by then.

It would be a pity to undervalue the achievements of the post-war piston-engined types. They re-established thriving

Grumman Widgeon amphibian used by Tourist Air Travel for sight-seeing tours in New Zealand

airline services throughout the world, and notably over the Atlantic and Pacific Oceans which had still been regarded with some temerity in 1939. They pioneered mass, low-cost air travel and, taking full advantage of new radio and radar aids, opened up completely new routes. Boldest of them was the 'over-the-Pole' route between Europe and North America.

Hitherto, it had usually been left to America, Britain, France, Germany and the Netherlands to set the pace; this time the lead was given by Scandinavian Airlines System (SAS), which linked the airlines of Denmark, Norway and Sweden. Beginning on 15–16 November 1954, it operated a trans-Polar service from Copenhagen to Los Angeles, via Sondre Stromfjord in Greenland and Winnipeg, Canada, using DC-6Bs. Following closely the ideal Great Circle route, it cut the overall journey time to twenty hours, compared with the former thirty hours via New York. Special navigation systems had to be evolved, because of the effect on standard compasses of flying near the magnetic pole; but weather proved less of a hazard than had been expected, and

The rear loading doors of the Fairchild Packet pioneered modern quick-loading methods for freight

soon Polar flying was regarded as no more unusual than any other kind.

In just a few places throughout the world there was still a use for the flying-boats which had been discarded by the major airlines. From the UK Aquila Airways flew holiday-makers to Madeira in Hythes and Solents until 1958. Even after that, one could still see the odd Short 'boat around islands in the Pacific and in the Argentine; but the only flying-boats in airline service today are found on local routes in places like New Zealand and on the short hop from San Pedro, California, to Catalina Island.

A Bristol 170 Freighter of Silver City's cross-Channel vehicle ferry fleet

ANYTHING . . . ANYWHERE

Freighting gathers momentum

When the last generation of piston-engined airliners was
retired from the main passenger routes, some operators had
the aircraft converted into freighters. Typical were the
DC-7Fs, produced by building freight-doors into the side of
the retired DC-7Bs and Cs and strengthening the cabin floor
to take concentrated loads. Such aircraft could make money
because they had already given good service as passenger-
liners and had repaid most of their initial cost. They were,
however, less versatile than real cargo-planes.

The idea of building aeroplanes in a way that enabled
bulky loads to be put aboard and off-loaded quickly and
easily goes back to the 'twenties, when the Gloster company
drew up a highly-prophetic design. It was for a transport in
which the whole rear fuselage hinged to one side, opening
up the complete cross-section of a roomy cabin for rapid
loading. Nobody wanted this kind of aircraft at the time.

The first aircraft to demonstrate the potential of future
freight-planes were Second World War gliders. After landing,
British Horsas could shed their rear fuselage, enabling men
and equipment to get into action quickly. Some really big
gliders, like the Hamilcar, had sideways-hinged noses, so
that tanks and troop-carriers could be driven straight out of
the cabin, down ramps. And a completely new concept was
pioneered by America's Laister-Kauffman XCG-10A, which
had clam-shell doors at the rear of its pod-like cabin, enabling
vehicles and equipment to be loaded and unloaded easily
under the high tail-boom.

Side-hinged noses and rear doors were to become standard

The DC-7F Speedfreighter was a converted DC-7B/C airliner

The long-nosed Superfreighter could carry three cars and twenty passengers. (*Below*) carrying racehorses by air is big business. Naturally nervous, the animals are less affected than by rough sea crossings.

features of post-war freight-planes; but the Fairchild C-82 Packet, first of the highly specialized freighters in 1944, embodied yet another alternative. Its cabin consisted of a roomy square-section pod, with the floor at truck-bed height, and with rear-loading clam-shell doors between twin tail-booms. Vehicles could back up to the wide-open cabin between the booms, to discharge their loads into the hold, or could themselves drive up ramps into the hold.

This method of loading, and the nose-doors on the later Bristol Freighter, revolutionized cargo-carrying by air. The US Army began to reshape whole divisions so that they could be carried, with their equipment, inside Packets. Paratroops could be dropped from side-doors in the cabin. If the rear doors were left off, heavy loads could be para-chuted from the back, helped on their way by roller conveyors in the floor.

The Packet was, in fact, so useful that it 'killed' itself as a design. Realizing that air-drops were as important as quick loading on the ground, the military authorities and manu-facturers switched from the clam-shell rear door/twin-boom idea to a variation of the Laister-Kauffman layout, with the rear of the cabin made in the form of an upswept floor-door, hinged at the forward edge so that it could double as a loading ramp (see page 145). By the time an overhead gantry had been built into the cabin ceiling, for easy handling and positioning of heavy loads, and a powered winch added at the front of the cabin for hauling them aboard, all the basic features of a modern air freighter had been worked out.

The operation which, more than anything, demonstrated the potential of air freighting was the Berlin Air Lift of 1948–49. For nearly a year, a city of $2\frac{1}{4}$ million people was kept alive and at work solely by using an armada of freight-planes to ferry in everything they needed. Altogether, 2,326,205 tons of supplies were flown into the beleaguered city in 277,728 sorties by US, British and French aircraft.

Commercial shippers could not fail to be impressed, and soon Silver City Airways were ferrying motor cars over the English Channel for businessmen and holidaymakers, at a fare that attracted tens of thousands of people away from the normal shipping services every year.

Canadair's swing-tail CL-44D4 freighter

A 'visor' nose and kneeling undercarriage facilitate loading of the huge C-5A Galaxy

Swing-tails and Guppies

The aircraft which first took advantage of the old Gloster 'swing-tail' idea on a big scale was the Canadair CL-44D4. Evolved in Canada from the Britannia airliner and powered by four 5,730-hp Rolls-Royce Tyne turboprops, it can carry a payload of up to $28\frac{1}{2}$ tons and fly 5,260 miles non-stop at 380 mph with smaller loads.

Even the CL-44D4 is small by comparison with some of the freighters now flying. For example, the 81 Lockheed C-5A Galaxies built for the USAF are the biggest and heaviest aeroplanes in the world. Each spans 222 feet $8\frac{1}{2}$ inches, is powered by four 41,000-pound-thrust turbofan engines, and can carry more than 118 tons of payload in its cavernous hold.

The nose of the Galaxy hinges upward, like the visor of a mediaeval knight's helmet, and its 28-wheel undercarriage can 'kneel' to facilitate loading. Maximum range with a 35-ton payload is 6,500 miles, at 537 mph; a full load can be flown 2,950 miles.

The Galaxy is, of course, a military aeroplane; but the last fifty years have shown over and over again how quickly the airlines take advantage of concepts pioneered by (and paid for by!) air forces. Already Lockheed are offering a commercial version of the same design, with more powerful engines and a take-off weight of $383\frac{1}{4}$ tons, including a $142\frac{1}{2}$-ton payload. This would enable car manufacturers to ship their products over the Atlantic, from 63 to 111 at a time in a single aircraft, depending on the size of the vehicles, and would work out less costly than present transport by ship.

Big as it is, there are some items that even the Galaxy cannot carry – like the S-IVB top stage of the Saturn V/Apollo rocket, which has to be flown from the factory where it is built to Cape Kennedy. It was to cope with such outsize loads that Aero Spacelines devised the Guppy family of transport aircraft. Each of these began life as a Stratocruiser airliner or its military equivalent, the C-97. The fuselage is lengthened and its whole top deck is replaced by an enormous new 'bubble' structure, up to twenty-five feet in diameter in the Super Guppy version, with the nose hinged to swing to one side for easy loading. It is in aircraft of this type, perhaps the ugliest in the air, that parts of the twentieth century's most exciting vehicle begin a journey that leads to the Moon.

Only the Super Guppy can fly the S-IVB stage of America's Saturn V Moon rocket

(*Left*) Bell Model 47 crop-dusting. (*Below*) largest helicopter in passenger service is Russia's Mil Mi-6.

Choppers in action

When Igor Sikorsky perfected the first entirely practical 'single-rotor' helicopter (pages 98–99), he could hardly have foreseen the great variety of jobs that descendants of the VS-300 would be doing within thirty years; but his company has continued to be a world leader in helicopter development right up to the present time.

As in the case of the freight-plane, the helicopter's capabilities have tended to be exploited first by the armed forces and then made available commercially. It was, for example, in Korea in 1950–53 that the 'chopper' (to give it its familiar

(*Below left*) Sikorsky UH-34 (S-58) retrieving a Mercury astronaut from the sea; (*right*) a Westland Wessex flying crane in action.

nickname) first became a fully-fledged weapon of war. As front-line transports, Sikorsky H-19s hauled squads of troops and their equipment speedily to almost inaccessible mountain positions. Equally important was the work of the rescue helicopters, which evacuated 22,000 wounded men, often under fire, and reduced the death rate from injuries to the lowest figure in any war up to that time.

Supply, troop dropping and casualty evacuation remain three of the main jobs for military 'choppers', on a steadily increasing scale. One of the end products of such requirements is the big 'flying crane'. It is hardly beautiful, often consisting of no more than a backbone structure with the power plant and main rotor mounted at the centre of gravity, a tail anti-torque and control rotor, a stalky undercarriage, a cab at the front for the crew, and a separate rearward-facing control cab under or to the rear of the main flight deck from where a pilot can control the helicopter and keep an eye on the payload at the same time during loading and unloading.

One such flying crane is the Sikorsky CH-54 (civil S-64), which can carry a ten-ton payload under its slim structure. A typical load is the Universal Military Pod, a square-section container which can be equipped to carry forty-five combat-ready troops or twenty-four stretchers, or can be fitted out as a field hospital, command post or communications centre. It all sounds very military; but the city of Los Angeles decided to follow up a good idea by building a similar kind of 'people pod' as a mobile airport lounge. This can be towed on the ground by a variety of vehicles, picking up airline

A Sikorsky CH-54A Skycrane delivering its Universal Military Pod

passengers from various points in the city, then taken to a heliport, from where it is ferried to the main airport by S-64 Skycrane, and finally towed out to an airliner. Passengers can thus get from city-centre to the cabin of their airliner without changing from one vehicle to another in the process.

The huge loads lifted by some of the latest helicopters offer all kinds of other possibilities. Flying cranes could be used for unloading ships where there are no harbour facilities. Already Russia uses its giant Mil Mi-6s to ferry oil drilling rigs to places in Siberia where winter conditions and the absence of roads and airfields hamper the use of other forms of transport. The biggest of the Mils, the Mi-12, has set up official records by lifting loads of nearly forty tons. This opens up all kinds of exciting and profitable possibilities in an aircraft that can go anywhere and lift objects of any shape because it does not have to put them inside a cabin.

It is only as a straightforward passenger transport that the 'chopper' has yet to find its place. Airlines like BEA and Sabena have done their best to make such services economically feasible; but helicopters have to pay for their versatility by being more costly to operate than a fixed-wing aircraft of the same size. They are operated by BEA between the mainland and the Scilly Isles, and by Aeroflot between Moscow and its airports, and to recreational centres on the Black Sea, as a service to the public. Such flights do not show a profit and are not expected to do so.

Helicopters can be used economically on jobs such as agricultural crop-spraying, because their ability to operate from any open space on the edge of the field to be treated keeps down the costs. On other occasions – when, for example, they haul exhausted swimmers out of the sea, evacuate injured climbers from a mountain ledge, or retrieve from the sea astronauts who have been to the Moon – the cost of the operation is less important than the ability to do it at all.

(*Opposite page, top left*) Westland Whirlwind rescuing a ditched pilot from his rubber dinghy; (*top right*) a Mil Mi-4 lowers crew members to a Soviet submarine; (*centre*) one of the Sikorsky S-61Ns used on BEA's Penzance-Scilly service; (*bottom*) the big Mil Mi-10 can even carry a motor coach on its payload platform.

Germany's Heinkel He 178, the first jet-plane to fly, on 27 August 1939

DROPPING THE PROP

The first jets

Jet propulsion was no new idea when it came into everyday use in aeroplanes. Nearly two thousand years ago the Alexandrian philosopher Hero devised a steam-jet device called the Aeolipile which is said to have been used to open the heavy doors of a temple. In 1867 two Englishmen named Butler and Evans patented a steam jet-propelled aeroplane with delta wings. And the Romanian designer Coanda exhibited a kind of jet-propelled biplane at the second *Salon de l'Aéronautique* in Paris in 1910.

However, the true story of the modern jet-engine, or turbojet, began in 1928, when a young cadet named Frank Whittle, at the RAF College, Cranwell, wrote a thesis entitled 'Future Developments in Aircraft Design'. At a time when the fastest RAF fighters flew at a mere 150 mph, he looked forward to when aeroplanes would travel at 500 mph, at heights where the air is much 'thinner' than at sea level. Piston-engines and propellers would never make this possible, and he envisaged the use of rockets or what we would now call a turboprop.

At first, the Air Ministry were not interested in Whittle's

From the pioneer Gloster E.28/39 and its Whittle engine were evolved the Rolls-Royce-powered Meteor jet-fighter

theories. So, with some friends, he formed a small company called Power Jets Ltd. Using a new nickel alloy named Nimonic, specially developed by Mond Nickel Company, Power Jets produced an engine that could withstand the great temperatures created inside the combustion chamber, and ran it for the first time on 12 April 1937. By mid-1939 an improved version was so promising that the Air Ministry asked the Gloster Company to design an aeroplane in which to flight test it.

Long before that prototype was built, the first flight of a jet-plane, the Heinkel He 178, took place in Germany, on 27 August 1939. Neither the He 178 nor its engine, designed by Pabst von Ohain, achieved much more than this historic 'first', but they inspired the German industry to begin serious work on jet-fighters, and this led eventually to production of the very fine Messerschmitt Me 262.

Italy was next to fly a jet-plane, the Caproni Campini N.1, but this was not very exciting as the fan which produced its jet was driven by an ordinary piston-engine. By comparison, when the Gloster-Whittle E.28/39 flew for the first time on 15 May 1941, Gloster already had in production a twin-jet fighter named the Meteor to take advantage of the new method of propulsion that Whittle had perfected.

Jets and turboprops

The first jet combat aircraft played only a small role in the Second World War; but their performance was so superior to that of piston-engined types that the days of propeller-driven fighters and bombers were clearly numbered.

When the time came to consider using jet-engines in commercial aircraft the picture was slightly different. There was a variety of 'jet' in which the gas-turbine was used primarily to turn a propeller, instead of simply to produce propulsion by jet-thrust. First tested on a Meteor fighter in September 1945, this type of engine was called a propeller-turbine, or turboprop. Being tied to a propeller it did not offer quite the same high speeds and high-altitude cruising as a pure turbojet, but could give a much higher overall performance than any piston-engine, combined with greater fuel economy than any contemporary turbojet. As a result, in the late 'forties, everyone believed that turboprops would be used for long-range airliners and those short-haul machines for which low operating costs were important, leaving turbojets to power the medium-range 'prestige' airliners to which passengers would be attracted by speed alone.

Britain's second-generation turboprop airliner, the Bristol Britannia

A Vickers Viscount of Lufthansa

The jet age in air transport dates from 27 July 1949, when the prototype de Havilland Comet flew for the first time. It was a spectacular success, and the first jet airliner service with fare-paying passengers opened on 2 May 1952, when the BOAC Comet 1 G-ALYP took off for Johannesburg.

BEA, too, had decided to take advantage of the faster, smoother flying 'above the weather' offered by turbine engines. In view of their shorter routes they went for turbo-props, and began operating Viscounts on 18 April 1953.

For BEA the gamble paid handsomely. Their Viscounts were the envy of all other airlines, and Vickers eventually sold a total of 444. It was the first time that any British commercial transport had achieved such success in the world market, but might have been rivalled by the Comet had not disaster overtaken the BOAC machines. Accidents in 1954 were traced to the little-known phenomenon of metal fatigue, resulting from the high pressurization needed for flight above 30,000 feet. Before the fault could be remedied in later Comets, Britain had lost its lead, with Russia's Tu-104 already in service and America's Boeing 707 and Douglas DC-8 soon to follow.

Russia's Tupolev Tu-104

The prototype de Havilland Comet 1

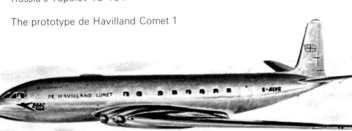

Boeing 707, first of the 'big jets'

Douglas's counterpart, the DC-8

No more propellers

Despite the tragic setback to the early Comets, the new
Comet 4 entered service with BOAC in time to operate the
world's first transatlantic commercial jet services on 4
October 1958. It was not really designed for the Atlantic run,
being a medium-range aircraft. The same was true of the
Boeing 707-121s of Pan American, which began operating
between New York and London on 26 October; but the true
intercontinental 707-320 and -420 'big jets' were not far
behind.

PanAm's initial order in 1955 had sparked off the biggest
airline spending spree of all time. With a price tag of around
£2½ million per aircraft, the president of one US airline proclaimed with feeling: 'We are buying aeroplanes that haven't
yet been fully designed, with millions of dollars we don't
have, and are going to operate them off airports that are too
small, in an air traffic control system that is too slow, and
we must fill them with more passengers than we have ever
carried before.'

We know now that his fears were unfounded. By the time

the 707 and DC-8 were in service, airfield runways had been extended to accommodate them. Air traffic control coped well with the higher speeds, and passengers came forward in greater numbers year by year to make the aircraft profitable. Boeing alone were destined to sell more than 2,000 jet airliners during the next fifteen years, including 200 aircraft far bigger than the 707, costing around £10 million each.

It took only a short period of jet operations to make the airlines realize that piston-engines were thoroughly outdated, even for freighting. The turboprop airliner was next to disappear from the trunk routes, as turbojets gradually became more economical and were joined eventually by turbofans, in which the turbines drive large-diameter front fans to give better fuel economy.

Today, airliners powered by turbofan engines dominate the world's long and medium stages. Small turboprop aircraft like the HS 748 and Fokker Friendship are still popular on short routes, but even here the turbofan is expected to take over eventually.

Fashions in airframe configuration have changed as much as engine design. The 'buried' engines of the pioneer Comet

Caravelle, first rear-engined jet-liner

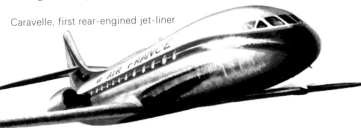

The four-engined BAC (Vickers) Super VC10

Tupolev medium-range twin-jet Tu-134

and Tupolev Tu-104 imposed serious limitations. Wing spars had to pass around them and it was not easy to accommodate newer, larger power plants. The underwing pods of the 707 and DC-8 appeared to offer many advantages over buried installations. Servicing and replacement were easier and larger engines could be substituted without too much redesign.

Then came the French Caravelle, with its two Rolls-Royce Avon turbojets mounted on each side of the rear fuselage. This left the wing uncluttered, enabling it to do its job of providing lift without interference, and the passenger cabin was quieter than ever before, because the jets were behind the people.

Not until later was it discovered that rear-engined jets had one inherent weakness. Most of them had a high T-tail, with the tailplane at the top of the fin. If they happened to stall – perhaps because the pilot throttled back too much in bad weather or while landing – the wing tended to 'blanket' the

The Trident's tri-jet layout inspired Tupolev's fine new Tu-154.

Another three-engined jet-liner: Boeing's very popular Model 727

tail surfaces and it was impossible to prevent the aircraft from falling to the ground. After this, the rear-engined jets had 'stick shakers' or 'stick movers' installed, to warn the pilot or even take over control before a 'deep stall' condition could threaten their safety.

There are several variations of the rear-engined configuration. The VC10 and Russian Il-62 each have four engines, in pairs on each side of the fuselage. The designers of the British Trident, on the other hand, settled for three engines, with one each side and the other in the tail, supplied with air by a duct at the base of the fin. Boeing followed this formula in their Model 727, as did Tupolev for the Tu-154, and Lockheed for their wide-bodied L-1011 TriStar airbus. On the DC-10 airbus, McDonnell Douglas have built the third engine into the fin, above the fuselage; but Boeing reverted to underwing pods on their big Model 747 'jumbo jet' and short-range twin-turbofan Model 737. Only long experience will show which arrangement is best.

Reversion to underwing pods on the Boeing 737 was a surprise

Jets for businessmen

The belief that jet-engines were too uneconomical for use in small civilian aircraft died many years ago. A French company named Turboméca fitted a tiny turbojet to a Sylphe powered sailplane back in the 'forties. It was purely experimental, but Turboméca went on to become the world's leading manufacturer of small gas-turbine engines for fixed-wing aircraft and helicopters.

The French seem to specialize in miniature 'jets', for another company, Microturbo, has supplied Eclair turbojets, rated at only 176 pounds of thrust, to power current sailplanes of French and Italian design. The idea is to enable such aircraft to be independent of winch launching or aero-tows, by taking off under their own power and then switching off the engines for gliding flight. The main snag is that the weight of engine and fuel detracts from the performance of the glider.

When we go up the size scale to the 'business jets' no such snags are apparent. Companies learned the value of providing fast door-to-door transportation for their highly-paid executives years ago, and there is nothing faster than these jet-liners in miniature.

First in the field was Lockheed's JetStar, which is unique in having four engines in pairs, like a VC10. It has remained in production for thirteen years, and continues to sell because of the very high-speed (570 mph) transportation that it offers for its eight or ten passengers.

Most business jets have twin engines, like the 7/12-passenger Hawker Siddeley 125, with 3,360-pound thrust Viper turbojets, or the ten-passenger French Dassault

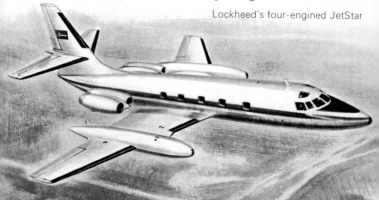

Lockheed's four-engined JetStar

The RAF uses a version of the Hawker Siddeley 125 for aircrew training, as the Dominie.

Mystère 20, with 4,250-pound thrust General Electric CF700 turbofans, which is marketed by PanAm in America as the Falcon. They are a little slower than the JetStar, but more economical to run and hence sell in greater numbers.

The standard of comfort offered by such aircraft is high. Cabins are often furnished to the owner's own specification, but reclining chairs are standard, with tables, wardrobe for coats, a buffet and cocktail cabinet. After that the sky is the limit, with stereo players and tape recorders as a starting point; but none goes quite so far as the larger executive aircraft ordered by a Middle East potentate, who travelled on a throne so mounted that it could always face Mecca no matter in which direction the aircraft was flying.

(*Below*) two very different 'baby jets' from France are the Dassault Mystère 20/Falcon business 'plane and the Fauvel AV.45-01R self-launching sailplane

Lightplanes by the thousand

American manufacturers of lightplanes and business aircraft alone sell more than 12,000 each year, with a total value of over $450 million. Nearly half of them are produced by a single manufacturer, Cessna, who offers about thirty different models, ranging from small 100-hp two-seat trainers to six/ten-seat executive transports with twin piston-engines of up to 375-hp each, and a twin-turbofan business jet.

Least conventional is the Cessna 337 Super Skymaster, which set a completely new fashion for twin-engined business aircraft when it first appeared in 1961. Up to then, machines in this category had tended to follow the same formula, with high or low monoplane wings, carrying an engine to each side of a conventional fuselage with enclosed cabin.

The average owner, lacking the experience and skill of an airline pilot, tended to feel in trouble if he suffered an engine failure and had to get down on a single engine that was trying to push the aeroplane round in circles. It was to offset this difficulty of flying with asymmetrical power that Cessna

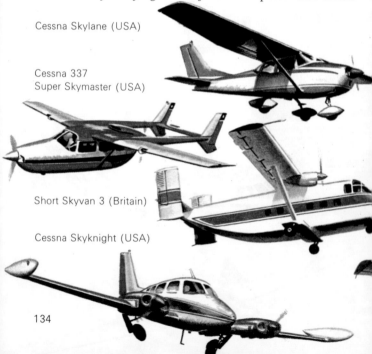

Cessna Skylane (USA)

Cessna 337
Super Skymaster (USA)

Short Skyvan 3 (Britain)

Cessna Skyknight (USA)

evolved the 'push-and-pull' Skymaster. By putting one engine in front of the cabin and the other at the rear, between twin tail-booms, if either engine should fail the other will continue to exert its thrust along the aircraft's centre-line. There are no unusual control problems, and the aircraft can maintain height with either engine shut down.

The same idea prompted the construction of the 'push-and-pull' Socata/Matra M360 Jupiter in France, but this has remained merely a promising prototype. In other respects, Socata is one of the small number of manufacturers outside the USA that has succeeded in competing with the American 'big three' of Beech, Cessna and Piper.

Key to Socata's initial success was that it took over a number of good basic designs produced by other small companies and gave them the sound backing of its parent, the state-owned Aérospatiale (then Sud-Aviation). Typical of these machines is the three/four-seat Morane-Saulnier Rallye, which comes in a range of models with engines from 100- to 220-hp. It is truly delightful to fly, with full-span

Beagle Pup (Britain)

Socata Rallye (France)

Bolkow BO 208C Junior (Germany)

PZL-104 Wilga (Poland)

CEA/Jodel Ambassadeur (France)

Ten-seaters: the Italian Piaggio P166B Portofino (*top*) and British Britten-Norman Islander

wing slats and large flaps to give it a short take-off, slow-flying ability that spells maximum safety for the private pilot. Stalling speed is under 60 mph with a full load, with a top speed of 121 to 165 mph, depending on the engine fitted.

Rallyes are ideal equipment for flying clubs, and the same is true of the two/four-seat Pup, produced by Britain's Beagle company before it went out of business in 1970. With the Pup and the twin-engined B.206 in production, and useful foreign orders for the Bulldog military trainer counterpart of the Pup, Beagle looked well set to re-establish a lightplane industry in the United Kingdom. Unfortunately, the costs of manufacturing any aeroplane in comparatively small numbers are high and Beagle's owners, the British government, were not prepared to subsidize it further.

In contrast, British companies have found a worldwide market for two machines in the larger light transport category. One of these is the ten-seat Islander, produced by Britten-Norman in the Isle of Wight. The idea was to build the simplest possible all-metal transport to replace types like the old de Havilland Rapide biplane. It had to be able to land and take off on small grass fields, and be cheap to buy and operate.

Size of the Islander was kept down by designing the cabin with no centre aisle between the pairs of seats. The backs of the seats fold forward, like those of a car, enabling passengers to gain access to any part of the cabin via three side doors.

The Short Skyvan 3, powered by two 715-hp AiResearch TPE 331 turboprops, is equally practical. It can carry nineteen passengers or twelve stretcher patients, but is normally used as a freighter with a two-ton payload. Like the Islander, it can go almost anywhere and has, on at least one occasion, flown in parts to repair an aircraft which tried unsuccessfully to use the short and rough landing strips that it takes in its stride.

Smaller twins : the Dornier Do 28 (*top*), Piper Twin Comanche and (*bottom*) Czechoslovakian Morava

The Grumman Ag-Cat (*top*) and Snipe Commander are typical US agricultural sprayplanes.

Spraying crops and fires

Crop spraying and dusting is big business everywhere, and many aircraft have been specially designed for the job. They differ widely in shape and size, but share a few basic features. The structure has to be simple and easy to keep clean when corrosive chemicals are being dropped. Undercarriages are fixed, because there is no time to retract the wheels during brief up-and-down flights, many times a day; in any case, the heavy retraction mechanism would reduce the weight of

Veteran Yak-12s have been the mainstay of Soviet air ambulance services

chemicals that can be carried. Cockpits are high, to give the pilot a good view during his flights a few feet above the ground, frequently in hilly or tree-dotted country. And the fuselage structure is often designed in such a way that it collapses outward progressively from the nose in a crash at slow forward speeds, to protect the pilot if he makes a mistake and hits the ground.

The chemicals dropped by agricultural aircraft and helicopters destroy insect pests and weeds, make barren soil productive, speed plant growth and defoliate cotton plants for easier picking. In Russia, aircraft also spray coal dust over fields in Kazakhstan, to absorb the early spring sunshine and use its warmth to melt the snow a little earlier. In the same area, rice is sown from the air.

A surprising variation of the same idea practised regularly in America is fish-planting. It may sound rough, but the Fish and Wildlife Service has proved that lakes can be restocked with baby trout quite safely by cascading them from an aircraft skimming the water at a fairly low speed. This particular government agency uses aircraft for an incredible variety of other jobs, ranging from chasing and shooting wolves to searching for poachers.

One other giant-size task that aircraft tackle is the war against forest fires. All over the globe light spotter 'planes keep constant watch for the wisp of smoke that betrays the start of such fires. The source of the smoke is then attacked

Canadair CL-215 water-bomber in action over a forest fire

Piper Super Cub skiplane, famous for its rescue work in the Swiss Alps

by water-bombers and by 'smoke-jumpers' parachuted from aeroplanes and helicopters, in an effort to prevent a small fire from growing into a forest-consuming monster.

It was discovered long ago that seaplanes and flying-boats are ideal for this work. In countries like Canada and France, forests and lakes or rivers are often side-by-side. This means that the water-bombers can taxi over the surface of the water, scooping it up into tanks in their hulls or floats. Without even stopping, they can take off, drop the water on to the fire and return for another load. A Canadian company

de Havilland Canada Otter floatplane, one of the DHC family of STOL transports

bought the US Navy's four Martin Mars flying-boats – then the largest in the world – for conversion to water-bombers when they were no longer needed for military service. Each of the big 'boats could carry and drop 8,000 gallons of water on a fire in a single flight. The results were so impressive that the Province of Quebec decided to buy from Canadair a whole fleet of CL-215 twin-engined amphibians specially designed as water-bombers. Although much smaller than a Mars, each can pick up 1,200 gallons of water in a 12-second taxi-run across a lake, and 'bomb' a fire 100 miles from its base 75 times in a single day, with only one stop for refuelling.

On work of this kind, in remote areas, the seaplanes and flying-boats are still unrivalled. The only exception is in winter, when rivers and lakes become frozen. Then it is the turn of the ski-plane to show its paces. Nowadays, 'wheel-skis' are fitted more often than plain skis. They are designed in such a way that they fit around the normal wheels of a landplane and can be retracted upward a few inches to permit normal landings and take-offs when there is no snow.

Ski-planes are used, with helicopters, in the Alps to pick up injured climbers whom it would sometimes take a mountain rescue team many hours, or even days, to reach. On a happier note, bigger STOL ski-planes, like the turboprop Twin Otter, are operated by Air Alpes into altiports atop mountains, to deliver holiday-makers quickly to leading ski resorts.

A wheel-ski-equipped Twin Otter transport of Air Alpes

Modern home-builts

The amateur flying movement of which Santos-Dumont and Henri Mignet had dreamed began to materialize in France soon after the Second World War, when Monsieurs Jean Delemontez and Edouard Joly formed a company to design and build a small, simple, wood and fabric single-seat aeroplane known as the Jodel D.9 Bébé. A 25-hp Poinsard engine gave it a reasonable performance, with a top speed of 94 mph and range of 250 miles at 71 mph. The French government realized that such aircraft offered enthusiasts an opportunity of getting airborne at low cost and devised a scheme to help anyone interested, by repaying much of the cost of a home-built lightplane once it had proved its ability to fly.

The French authorities also agreed to cover the cost of developing a two-seat version, the D.11, and eventually Jodel marketed kits of both the Bébé and its larger brother. Those people unable to afford a kit of parts could buy a set of plans for only £10 and search for their own raw materials.

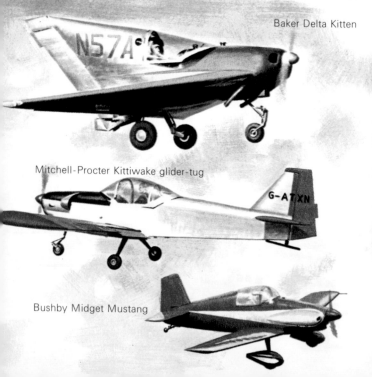

Baker Delta Kitten

Mitchell-Procter Kittiwake glider-tug

Bushby Midget Mustang

Aircraft were built in garages and bedrooms, often over a long period if their constructors could afford only a modest quantity of wood at any one time.

It was much the same story in the States, where the Experimental Aircraft Association was formed to co-ordinate the efforts of the home-builders, and represent them in dealings with the government agencies controlling flying. A big concession, now applied almost everywhere, is that the amateur-built machines do not need a costly annual Certificate of Airworthiness so long as their construction is supervised by experts and they are flown in accordance with strict rules that, for example, prohibit flying for hire or reward and after dark.

Today, thousands of 'home-builts' are giving their owners safe, cheap flying in every corner of the globe. Most of them are far more advanced than the original Jodels. Two or three of them have even been put into large-scale production by big manufacturers.

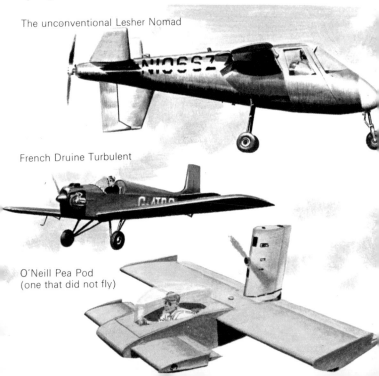

The unconventional Lesher Nomad

French Druine Turbulent

O'Neill Pea Pod
(one that did not fly)

Air power for peace

The Berlin Air Lift of 1948–49 (page 117) did more than simply keep $2\frac{1}{4}$ million people alive while politicians argued. It showed for the first time that the transport aeroplane could play as big a role in the cold war between east and west as the atomic bombers which prevented an all-out shooting war by the threat of mass annihilation.

From that moment, air forces paid fresh heed to building up strong transport commands, equipped with aircraft that not only carried huge loads but could put them down almost anywhere and drop them by parachute where a landing was completely impossible.

As already explained, helicopters can take care of many transport tasks in combat areas and where distances are short. They have their rivals in small STOL transports like the Russian Antonov An-14 Pchelka (Little Bee), which can take off with a full load of seven to nine persons in only 100 yards and land in 120 yards. And, of course, for long-distance transport, there is still no alternative to the big

The little Antonov An-14 eight/nine-seater is in both civil and military service

Heaviest aeroplane yet flown is the USAF's Lockheed C-5A Galaxy transport.

fixed-wing aeroplane with its great carrying capacity.

There is one type of military operation for which turbo-prop transports still have much to offer. Very high speed is often less critical than the ability to carry a heavy load at around 350 mph and put it down on a comparatively short, unprepared airstrip. That explains why Lockheed's C-130 Hercules, with four 4,050/4,500-hp Allison T56 turboprops, has remained in production for nearly twenty years, during which time well over a thousand have been built for many air forces and a number of freight airlines.

Payloads of up to twenty tons can be hauled by the Hercules. They are loaded via a ramp-door at the rear of the cabin, and this can be opened in flight to permit the air-dropping of freight by parachute. Britain's turboprop Short Belfast is an even bigger aircraft on the same lines; but here, as in the commercial airliner field, the turbofan engine is beginning to supersede the turboprop, and the biggest military transports are already turning to this type of power plant, which can offer both high speed and fuel economy.

Airdropping freight from a Lockheed Hercules of the RAF.

The Short Belfast can carry a 35-ton payload at 350 mph.

Giant turboprop transports from Russia : the Tu-114 (*top*) and An-22

LARGER AND FASTER

Enter the Jumbo

It took the Boeing 707 and Douglas DC-8 'big jets' only a few years to prove that they were the most profitable and safe airliners ever put into service, and from that moment the airlines knew they would soon be flying even bigger aircraft.

The economists had no difficulty in proving that it costs less to carry 300 people in one big aeroplane than in two 150-seaters. Air traffic controllers, who were snowed under with arrivals and departures at big airports, also liked the idea of using fewer aircraft to carry a particular number of people. Only the airport authorities were slightly worried at the thought of coping quickly with hundreds of persons arriving on a single aeroplane, and the mountains of baggage they would bring with them.

It was, therefore, inevitable that the 'big jets' would soon be joined by bigger jets, super jets and what the press began to refer to as 'jumbo jets'. Russia had already maintained its reputation for giantism with the Tu-114, produced in time to mark the fortieth anniversary of the Communist revolution, in 1957. It was a great aeroplane in every way, able to carry up to 220 passengers, or fly a smaller number on routes from Moscow to places like Havana, Cuba. It was so big that its huge restaurant did not need to be occupied except at meal-times, and the kitchens were on a lower deck, staffed by genuine tall-hatted chefs.

Even the Tu-114 was far exceeded in size by the Antonov An-22 which Russia revealed in 1965; but by this time the turboprop was outdated for long-range passenger travel and a projected 724-seat 'airbus' version of the An-22 was dropped. It was left to America to set the pace in true 'jumbos', first with the enormously-stretched 259-seat McDonnell Douglas DC-8 Super 60 family and then with the 308/490-seat Boeing 747. This has a cabin twenty feet wide and eight feet high, making it the pace-setter of a completely new generation of huge wide-bodied transports which will carry much of the world's airline traffic in the 'seventies.

The McDonnell Douglas 'stretched' DC-8 Super 61

Boeing 747, first of the 'jumbos'

Dornier Do 31

Fairey Rotodyne

LTV-Hiller-Ryan XC-142A

A Westland project for a tilt-wing VTOL airliner

VTOL

The helicopter will never be bettered for many jobs that require an ability to take off and land vertically. It is not exactly quiet; but most of the alternatives so far offered are so much noisier that it would be impossible to operate them into places like city-centres without upsetting householders and workers for miles around. This has led to one of the most intensive research programmes in aviation history.

So far, most experiments in finding VTOL (vertical take-off and landing) alternatives to the helicopter have been paid for by the military, who would like to find a way of combining the 'chopper's' go-anywhere versatility with the speed of a fixed-wing aircraft. America, in particular, has tested an incredible variety of designs, with tilting wings, tilting rotors, tilting propellers inside ducts, swivelling engines, huge wing flaps to deflect the slipstream of turbo-props, lift-fans housed inside wings, vertically-mounted jet engines and . . . you name it, they've tried it!

Best of the lot was probably the big XC-142A, which had four turboprops mounted on a tilting wing. The turboprops worked like helicopter rotors for take-off. At a safe height, the wing was tilted down – a horizontal tail rotor helping to stabilize the aircraft at this time – and the XC-142A then cruised like a conventional aeroplane. This concept worked reasonably well and is still favoured by many designers, although some, like those of the Westland company, would exchange the propellers for rotors.

Another aircraft which worked well was Britain's Fairey Rotodyne convertiplane. This took off like a helicopter, with its big rotor driven by blade-tip jets. It cruised as an auto-gyro, with the rotor free-wheeling and the forward thrust coming from two wing-mounted turboprops.

The Rotodyne was abandoned because further development seemed too costly. Another programme that has ended is that for Germany's Dornier Do 31, which used Rolls-Royce lift-jets and turbofans with rotating exhaust nozzles to thrust it into the air, when the turbofans took over for high-speed cruising. The Do 31 was excruciatingly noisy; but one day one of these concepts, or something like it, must make possible the most important new aeroplane of its time.

Under construction in the early 'seventies, the Boeing 2707-300 will cruise at 1,800 mph

The passenger goes supersonic

It may be a long time before passengers fly in fast vertical take-off airliners, but they should begin to experience another major advance – supersonic flying – by 1973–74. This represents a tremendous achievement for the aircraft industry, because many people doubted that aeroplanes could fly faster than the speed of sound when the Second World War ended in 1945.

Wartime fighters had been able to dive fast enough for the airflow over their wings to be speeded up to near Mach 1*; when this happened, they were buffeted by shock-waves so intense that pilots sometimes lost control and aircraft broke up in the air. Not until Captain Charles Yeager of the USAF

* The term used to represent the speed of sound at any particular height: equivalent to 760 mph at sea level, falling to about 660 mph at 36,000 feet and above.

Russia's Tu-144, first supersonic transport to fly

The Concorde, landing with its nose drooped to improve the pilot's view

flew the specially-built Bell X-1 rocket-plane safely through the 'sound barrier' on 14 October 1947 could anyone be absolutely certain that supersonic flight was practicable.

Even then, it was some years before designers learned enough about the new aerodynamics needed for flight above Mach 1 to make this a safe, everyday routine for the pilots of jet fighters and bombers.

To the public, anyone who flew at such speeds was a superman. They pictured him dressed in a pressure-suit and 'goldfish-bowl' helmet, having to pass strict 'medicals' and subjected to face-distorting forces like the space-men who were still part of science-fiction.

Today, space-men have been to the Moon and the public can see in the air prototypes of aircraft intended to carry ordinary men and women, too, beyond Mach 1 – not just *at* the speed of sound, but at more than twice that speed. First

The Martin space transporter, used like a rocket to boost other craft into orbit, would probably use rocket engines for take-off and

to fly, on the last day of 1960, was Russia's Tu-144, designed to carry 100 to 121 passengers at around 1,550 mph (Mach 2.35). The Anglo-French Concorde, which followed it into the air in 1969, carries more people (up to 128) but is not quite so fast, with cruising speeds in the 1,350–1,450 mph bracket. This is fast enough to cut the London–New York time to a mere three hours, so that passengers will arrive before they take off in terms of local times; but America's Boeing 2707-300 will be even faster, with a designed cruising speed of 1,800 mph, carrying twice as many passengers.

Take-off of a typical space transporter, designed in Germany

turbofans for the return journey to its airfield, within Earth's atmosphere

Beyond the SST

Even to make their 1,800-mph SST (supersonic transport) possible, Boeing designers are having to use new materials like titanium and stainless steel for the airframe. This is not too much of a gamble, as considerable experience has been gained with research aeroplanes and a few military aircraft made of such metals.

Already America and several other countries have announced that they will not permit airliners to fly at supersonic speed over their territory until the boom has been eliminated.

External burning of its fuel gives this 5,000-mph airliner project a frightening appearance

Nobody is yet sure that this can be done, although scientists are working on it. Meanwhile, some designers are already thinking far beyond Mach 2 or 2.7, the speed of the Boeing 2707-300, to the sort of fantastic cruising speeds they believe will be within reach before the end of the present century.

If passengers are worried at the thought of faster-than-sound flight in the Concorde, it is fair to expect that they will be petrified even at the suggestion of what might follow. The space transporter proposed by the German industry some years ago was envisaged mainly as a freighter, or a means of getting astronaut crews to orbiting space-stations, by using a big rocket-plane to boost a smaller one into orbit piggy-back style. But the fearsome looking, overgrown 'paper dart' at the bottom of page 153 is a serious proposal for an airliner.

Wind-tunnel tests in Britain have proved that the strange inverted-W shape of its delta wing would be ideal for flight at around 5,000 mph – nearly eight times the speed of sound. Climb and acceleration to cruising speed could be left to ramjet engines of the kind already used in many guided missiles; but when the aircraft reached its cruising height there would be a big advantage in switching to a form of external combustion, with the fuel actually being burned on the *outside* of the aircraft. How would passengers feel about travelling in an aircraft shrouded by intense flames? It may be as well to explain that they would not see the flames, which would be well behind the passenger cabin. In any case, they are quite accustomed to the thought of astronauts returning to Earth in a white-hot craft and would probably be prepared to accept the 5,000-mph airliner once they became convinced of its integrity.

This is just as well, for some designers are way ahead of even the external-burning, 5,000-mph airliner. Engineers at McDonnell Douglas have put forward ideas for a rocket-liner called Hyperion which they claim could carry passengers, 170 at a time, from one side of the globe to the other in just forty-five minutes by the 'eighties. A trip from Los Angeles to Honolulu would take eighteen minutes.

You don't believe it will ever happen? There was a time when wise men proclaimed that if God had intended men to fly He would have given them wings!

The McDonnell Douglas Hyperion rocket transport has been put forward as a serious possibility for the 'eighties

One day, perhaps, wingless transports like this will ferry holiday-makers to huge orbiting space stations

BOOKS TO READ

L'Homme, L'Air et L'Espace by Charles Dollfus, Henry Beaubois and Camille Rougeron; Editions de l'Illustration, Paris, 1965.

A History of the World's Airlines, by R. E. G. Davies; Oxford University Press, London, 1964.

Aviation, an historical survey from its origins to the end of World War II, by Charles H. Gibbs-Smith; Her Majesty's Stationery Office, London, 1970.

The Great Planes by James Gilbert; Paul Hamlyn, London, 1970.

Warplanes of the Third Reich by William Green; Macdonald, London, 1970.

British Civil Aircraft 1919-59 (two volumes) by A. J. Jackson; Putnam & Company Ltd, London, 1959-60.

Kitty Hawk to Concorde by H. F. King; Jane's Yearbooks, London, 1970.

The Pocket Encyclopaedia of World Aircraft in Colour (series) by Kenneth Munson; Blandford Press, London, 1966-70.

British Aviation: The Pioneer Years and *British Aviation: The Great War and Armistice* by Harold Penrose; Putnam & Company Ltd, London, 1967/69.

European Transport Aircraft since 1910 and *Soviet Transport Aircraft since 1945* by John Stroud; Putnam & Company Ltd, London, 1966/68.

Aircraft Aircraft by John W. R. Taylor; Paul Hamlyn, London, 1970.

Combat Aircraft of the World edited by John W. R. Taylor; Michael Joseph, London, 1969.

Jane's All the World's Aircraft edited by John W. R. Taylor; Sampson Low, Marston & Co. Ltd, London, annually.

Jet Planes Work Like This by John W. R. Taylor; Phoenix House, London, 1969.

Pictorial History of the Royal Air Force (three volumes) by John W. R. Taylor and Philip Moyes; Ian Allan, Shepperton, 1968/70.

The Spirit of St Louis by Charles A. Lindbergh; Charles Scribner's Sons, New York, 1953.

INDEX

SOME OTHER TITLES IN THIS SERIES

Natural History

The Animal Kingdom	Butterflies
Animals of Australia & New Zealand	Evolution of Life
Animals of Southern Asia	Fishes of the World
Bird Behaviour	Fossil Man
Birds of Prey	A Guide to the Seashore

Gardening

Chrysanthemums	Garden Shrubs
Garden Flowers	House Plants

Popular Science

Astronomy	The Earth
Atomic Energy	Electricity
Computers at Work	Electronics

Arts

Architecture	Glass
Clocks and Watches	Jewellery

General Information

Arms and Armour	Guns
Coins and Medals	Military Uniforms
Flags	Rockets and Missiles

Domestic Animals & Pets

Budgerigars	Dogs
Cats	Horses and Ponies
Dog Care	

Domestic Science

Flower Arranging

History & Mythology

Archaeology	Discovery of
Discovery of	Japan
Africa	North America
Australia	South America
	The American West